# Methodology of Geophysical Research in Archaeology

## Vladimír Hašek

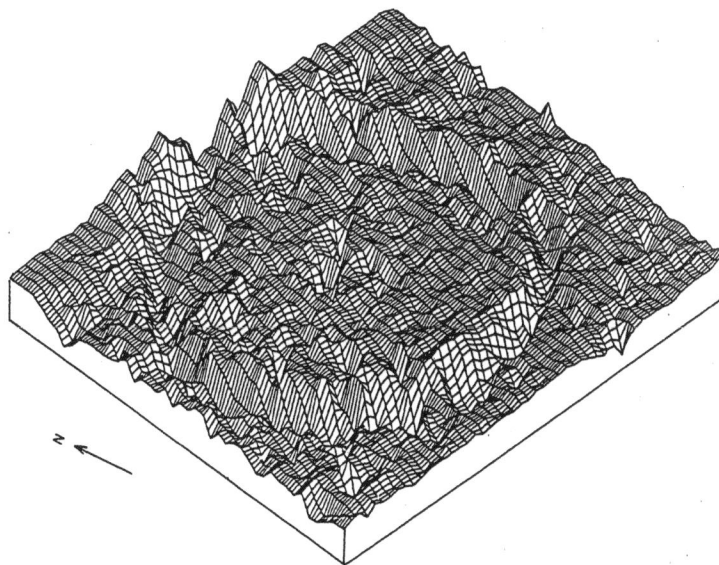

BAR International Series 769
1999

Published in 2016 by
BAR Publishing, Oxford

BAR International Series 769

*Methodology of Geophysical Research in Archaeology*

ISBN 978 0 86054 981 9

BAR Publishing is the trading name of British Archaeological Reports (Oxford) Ltd.
British Archaeological Reports was first incorporated in 1974 to publish the BAR
Series, International and British. In 1992 Hadrian Books Ltd became part of the BAR
group. This volume was originally published by Archaeopress in conjunction with
British Archaeological Reports (Oxford) Ltd / Hadrian Books Ltd, the Series principal
publisher, in 1999. This present volume is published by BAR Publishing, 2016.

Printed in England

# BAR
PUBLISHING

BAR titles are available from:

BAR Publishing
122 Banbury Rd, Oxford, OX2 7BP, UK
EMAIL info@barpublishing.com
PHONE +44 (0)1865 310431
FAX +44 (0)1865 316916
www.barpublishing.com

Archaeology in practice is in essence the application of scientific methods in excavating old artifacts. It is based on the theory that the historical value of a certain object does not depend on itself but rather on further circumstances which can only be revealed by a professionally performed research.

Sir Charles Leonard Woolley
(1880 – 1960)

# CONTENTS

# LIST OF FIGURES

# LIST OF PLATES

# 1. INTRODUCTION

Archaeology is a science dealing with the oldest history of mankind on the basis of preserved material sources, irrespective of whether the remains of extinct cultures were discovered in excavations or whether they were preserved on the earth surface.

In archaeological practice there are several working stages (Podborský 1979: 34-45)

a) prospection in the field - oriented on finding a certain relic in the field by gathering of surface material, ground works, a complex of other methods, etc.,

b) archaeological investigation in the field - a process of obtaining valuable archaeological sources, which includes systematic, preventive, rescue, revision and other kinds of investigation,

c) restoration and preservation of sources of the material culture - above all laboratory work (i.e. the employment of nonprospective methods),

d) analysis and systematization of archaeological sources - (technical, natural historical, archaeo-logical-historical, etc.).

The above brief general overview points to the fact that due to its specific conditions and the relation of field prospection and investigation as certain methods for obtaining relics of the material culture, their evaluation and interpretation which, by broader conclusions about historical processes and the development of the human society in the oldest period, is the objective of prehistory, classical archaeology, medieval archaeology as well as further specific fields (Egyptology, industrial archaeology, postmedieval archaeology, etc.). This historical field needs perhaps most of all scientific disciplines in all its research stages a wide interdisciplinary co-operation, particularly with natural historical fields. It cannot do without anthropology, palaeozoology, palaeobotany and palynology, petrography, pedology, geology, mathematical and statistical methods, dendrochronology and a wide range of further dating methods (such as $C^{14}$, thermoluminiscence, knowledge of metallography, chemical analysis and neutronography). Even further technical fields, aerial prospection, the study of geodetic photographs, methods of experiment and others do not stand aside. Methods of archaeological prospection and the environment itself from which the monuments of the material culture are obtained are characterized by a close relation to geology, which is valid particularly for the earliest period - the Palaeolithic, but also for younger periods. It also means taking over and utilizing some methods and processes of geologic sciences, such as stratigraphy and lithology, without which the modern and comprehensive archaeological investigation cannot be conceived (Hašek - Měřínský et al. 1991: 9; Malina 1976: 43-67, 177-183).

That is the reason why the inclusion of geophysics into the set of methods applied in prospection, the investigation as well as in its evaluation was only a logical consequence of the further development of field archaeological practice in the Czech Republic, which is also in accordance with the present-day universal trend. Thus, as one of the tasks of applied geophysics is finding inhomogeneities due to the operation of different geological factors in the rock environment (Mareš et al. 1983; Müller - Okál - Hofrichterová 1985; Hašek - Uhlík 1991, etc.), it is possible, in the case of its utilization for archaeological purposes, to reveal these inhomogeneities in the form of different archaeological objects or structures created by man, i.e. remains of his activity during the development of human society. Geophysical prospection in archaeology can be defined as a process whose objective is the determination or precisioning of the frame groundplan scheme of the site studied or of the hitherto uninvestigated archaeological object on the basis of the combination of the physical model and accessible information collected by means of the surface investigation , sounding and further branches and methods of science (the study of written and iconographic sources, map material, aerial photos, etc.).

As a part of extensive investigation activity in systematic archaeological investigations and rescue activities, in the building of new communications, industrial objects, reconstruction of city centres, building of new urban complexes etc., the most effective ways were being examined

1) for finding out the localities and their prospection that would precede the excavation works,

2) for finding out the most suitable places and characterizing objects for detailed archaeological investigation,

3) for allowing to draw the archaeologist's attention to the choice of methods of the field archaeological investigation.

As optimal for the complex solution of these issues within both basic and applied investigation the interdisciplinary co-operation between geophysical institutions (Geofyzika a.s., the Faculty of Science, Charles University, Mining University in Ostrava) and archaeological ones (Archaeological Institutes of the Academy of Science, Czech Republic (further AS CR) Brno, Prague, Faculty of Arts, Masaryk University, Brno, the Czechoslovak Egyptological Institute, Charles University, central and regional museums, etc.) in utilizing different geophysical methods, particularly geoelectrical ones and magnetometry for finding archaeological objects in the field and further interpretations have proved in recent years in the Czech Lands.

This book, about the application of selected geophysical methods, their processing and interpretation on PC in archaeological prospection, sums up, generalizes and comprehensively evaluates results of more than a twenty-year-activity of authors in this field which has enriched the knowledge by a number of archaeological localities in the Czech Republic (Hašek, Měřínský1987, 1989; Hašek, Měřínský1991) and abroad (Hašek - Obr-Přichystal - Verner

1986; Hašek - Obr - Verner 1988; Verner - Hašek 1988; Fuchs - Hašek - Přichystal 1995; Hašek - Rössler-Köhler 1996; Hašek - Fuchs - Unger 1996; Hašek - Unger 1994 etc.), representing all periods of prehistoric, protohistoric and medieval historic development. Thus, an extensive interdisciplinary co-operation has been established between geophysics and archaeology, at whose establishment participated above all Prof. PhDr. J. Poulík, DrSc., the former head of the Archaeological Institute of the Czechoslovak Academy of Sciences, Brno, and the present head of the Archaeological Institute, AS CR Brno, PhDr. J. Tejral, DrSc, as well as Dr. B. Beránek, DrSc, of the state enterprise Geofyzika Brno.

This fundamental organizational structure gradually included a number of further fields of the natural historical sciences, technical and social sciences. In the past years in Moravia a purposeful combination of aerial prospecting with subsequent geophysical works followed by archaeological probing was applied (Bálek - Hašek - Měřínský - Segeth 1986; Bálek - Hašek 1986, 1996; Hašek - Kovárník 1996, etc.).

The co-operation has brought not only a large amount of new information (data, findings) and methods for archaeology itself on a qualitatively high level, but in partial findings the application of different methods of prospection and/or analyses it is also advantageous for the co-operating disciplines of natural historical and technical sciences.

The submitted information obtained in investigation and prospecting activities at a number of Moravian localities and abroad should serve archaeologists and specialists of related disciplines dealing with issues of asserting geophysical methods in archaeological investigation in the field and/or further persons interested who come in close contact with the results of these works. Therefore the possibilities of application of the geophysical methods for the above objectives are adapted for this purpose. The employed methods are listed as well as their main principles, factors limiting a broader application at the individual localities, methods of processing, interpretation, and it is hinted for what purpose and for the solution of what task the respective method can be of help.

On a number of examples in the location of objects of different character and from different periods of time the co-operation of social and natural historical branches for obtaining the maximum amount of knowledge and information from field investigation at the minimum expenditure of financial, material means and with maximum saving of time is documented.

Theoretical and practical experience and the results obtained and summarized in this paper/study/treatise contribute thus not only to the enrichment of archaeological investigation methods in the field, but also to the completion of the knowledge of the development of the human society from the Palaeolithic to the Slavonic period and the culminating Middle Ages.

All tasks included in this paper were fulfilled in co-operation with a number of colleagues from several archaeological and other institutions. Above all we would like to mention the geophysicists Dr. S. Mayer, Ing. R. Záhora,CSc., the programmers Mgr. J. Tomešek, Ing. R. Vencálek, Dr.J. Eisler, Dr. K. Segeth, CSc., the technician L. Švandová, the geologists Dr. F. Obr, Dr. A. Zeman, CSc., the archaeologists Dr. K. Ludikovský, CSc., Dr. R. Krajíc, CSc., Prof. Dr. B. Dostál, DrSc., Dr. J. Kovárník, CSc., doc. dr. Z. Měřínský, CSc., Dr. J. Peška, Dr. J. Unger, CSc., Prof. Dr. M. Verner, DrSc., Dr. M. Tymonová, Dr. J. Vignatiová, CSc., Prof. Dr. U. Rössler-Köhler, Dr. G. Fuchs, and a number of further persons without whom the present paper could not have been written and to whom we are much obliged.

## 2. ARCHAEOGEOPHYSICAL INVESTIGATION IN THE CZECH REPUBLIC AND ITS DEVELOPMENT

The first steps in the field of a broader use of geophysical methods in Bohemia were taken as early as in the early 1960s (Mašín - Válek 1963: 293-294; Linington 1969: 131-138). In the 1970s co-operation was started between the Department of Applied Geophysics, Faculty of Science, Charles University, Archaeological Institute of the CSAS in Prague and the state enterprise Geofyzika, enterprise Prague (Marek - Plesl 1978; Marek - Pleslová - Štiková 1977; Bárta 1971, 1973, etc.).

Archaeologists and geophysicists started dealing with the possibilities of employing geophysical methods in archaeology as early as in 1970. It was above all a broadly conceived comprehensive investigation of the extinct medieval village of Záblacany near Uherské Hradiště, where geoelectric methods were used for those purposes for the first time (Bárta 1971) and further works were carried out near the castle of Veveří near Brno (Bernat-Hašek 1973). The methods of resistance profiling, VES, the shallow refraction seismism and the drilling investigation were experimentally employed for solving the issue of locating the assumed medieval cavities and other near surface inhomogeneities in the area of the extinct medieval village. Further works took place only in 1973-1974, when, after obtaining proton magnetometers, the first extensive measurement at the well-known Great Moravian locality "Sady" near Uherské Hradiště was done (Hašek et al. 1975). In 1975 it was mainly the initiative of Dr. K. Ludikovský, CSc., (+) from the former AI CSAS in Brno who started, in co-operation with the authors of this paper/study/treatise, organizing and implementing systematic prospection of metallurgical installations in the Boskovice region (Ludikovský - Souchopová - Hašek 1977; Ludikovský 1978).

The overall conception of all investigation works in Bohemia and Moravia resulted in the creation of the IRB (Interdisciplinary Improvement Team) whose task was above all the study of the possibilities of application of geophysical methods in archaeology and their use in practice particularly for field prospection. The team, first of its kind in the former Czechoslovakia, commenced its activity in 1976. The Archaeological Institute in Brno and Prague, Geofyzika state enterprise, the Faculty of Science, Charles University in Prague and of further institutions, ministerial research formations, universities (University of Mining in Ostrava, Faculty of Arts, Masaryk University in Brno, etc.) joined their efforts to solve those issues. The main objective of the structure formed in this way was to remove some ministerial obstacles/barriers and to contribute to the safeguarding of the comprehensiveness, high effectivity, maximum economy of archaeological prospection and investigation and new views of the issues studied of the development of the human society from the Palaeolithic up to the present, bringing qualitatively and quantitatively more essential knowledge and methodological procedures which cannot be obtained by currently used investigation methods (Hašek - Ludikovský 1977: 117-119).

In 1976-96 these tasks already developed systematically in two main directions - in the sphere of field archaeological prospection and that of laboratory measurements.

The field archaeological prospecting followed two complementary and inseparable objectives of fundamental and applied investigation. The methodological-investigation works by means of geophysical methods in the field started from the knowledge obtained by surface investigations and collections, by preceding archaeological excavations at the localities studied and/or by/with aerial prospection and photography. The purpose was to construct maps of the measured magnitudes, later by means of computer technology (such as to draw shadow maps of $\Delta T$ anomalies, further depicting ways, spatial models, etc.) and to derive their geophysical interpretation. This method provided/granted the archaeologist the maximum amount of information either before the excavation works or parallel to the investigation, when, from the confrontation of both types of results methodological conclusions could be judged and possibly refined by further probing. What followed was on the one hand applied investigation, on the other hand the investigation-methodological aspect according to the character of geophysical methods and the existing instrumental technology. In the end the results became a guideline for a further practical application of geophysical methods in this line of science.

From the data and the results obtained, the employment of new instruments was studied as well. The objective was to create a situation permitting in the shortest possible way the implementation of applied investigation, the processing of the measured data and their representation in a graphic form, thus granting a clear basis for field excavations (Hašek - Ludikovský 1977: 117-119).

Laboratory measurements included the determination of physical properties of rocks, further auxiliary measurements, analyses, etc., thus suitably completing geophysical works in the field. They analysed, however, not only the measured data, but also the finds/findings obtained by the investigation and further information. Thus they were partly linked up with the prospection and with the analysis and interpretation of the data obtained and with archaeological information.

By a broader introduction of the rational complex of geophysical methods and aerial photography into archaeological and prospecting works (long-term, systematic, rescue, advance, finding ones, etc.) the following points were achieved:

1) a quick orientation in the field studied with the possibility of the choice of optimum areas for excavation works, the delimitation of protected regions, etc.,
2) a rational utilization of working capacities, economic costs and time savings,

3) obtaining more detailed data the archaeologist cannot reach without extensive time-consuming excavations, but which are essential for a deeper knowledge of the structures studied.

Field and laboratory measurements made it possible to compile and further refine the physico-archaeological model of the given field before the investigation. Linking up with it gradually further analyses were performed after finishing the field works whose results conditioned the full evaluation of the material and information obtained (such as Hašek - Ludikovský - Obr 1979; Hašek - Horák - Obr 1979; Hašek - Obr - Verner 1988; Hašek - Fuchs - Přichystal 1995).

From among the most important activities in the past twenty years in this field and results obtained in the co-operation of geophysics, geology and archaeology it is possible to mention a number of extensive and smaller measurements at open settlements, fortified localities, investigations of building relics, architecture, burial grounds and production centres.

In solving these main tasks in Moravia, above all proton magnetometry asserted itself. By a smaller scale of works geoelectric methods were represented (resistance profiling at different electrode arrangement, vertical electric sounding, later dipole electromagnetic profiling, georadar, etc.), shallow refraction seismicity and microgravimetry. Thus, in the years 1974-95 altogether 161 localities with the area of about 250 hectares were measured geophysically in solving different tasks, stripping works being so far carried out at only 92 places.

From the above material it follows that the highest volume of geophysical works was concentrated above all on the solution of issues of fortified localities and settlements (44), open settlements (40) and Neolithic circular sites (17). The ratio of geophysics and excavation work is approximately 5 : 3 (57.1 %).

A significant role in formulating the needs of archaeogeophysical prospecting and determination of further theoretical tasks and field works was played by meetings, conferences and symposia held on the territory of the former Czechoslovakia and in the Czech Republic regularly since 1973 (Hašek - Měřínský -Unger - Vignatiová 1983; Hašek - Měřínský - Págo 1983, 1984, 1985, 1987, 1987a; Hašek - Měřínský1987, 1987a, 1989; Hašek 1994, 1995, 1996).

# 3. METHODOLOGY OF GEOPHYSICAL WORKS

As follows from the introductory chapter, the objective of this paper is the summarization of the results of the authors' work from the field of methodology of measurement, processing and interpretation of data in magnetometry and geoelectric methods for the purposes of archaeological prospection and investigation.

Main stress is laid on:

1) the presentation of a uniform explanation and analysis of geophysical measurements for the investigation of objects of different characteristics and cultural appurtenance mainly from the area of central Europe, particularly Moravia and Silesia, but also Bohemia, Austria, Germany, and Ancient Egypt,

2) new forms of processing and interpretation of field data for the solution of the individual archaeological tasks,

3) methodological-investigation activity and field measurements for solving different tasks according to the demands of the individual archaeologists of the interested institutions.

The issues of the first part were prevailingly oriented on the optimization of geophysical works in the investigation of the following archaeological objects:

A) Settlement Monuments

a) open and fortified settlements,
- extent of settlement (cultural layers and objects),
- position, outer shape and size of the individual objects (settlement objects, further objects of the habitation, fortifications, etc.),
- properties of objects (a pot-house, a house with a stone footing, clay pit, stock pits, etc.),
- character of the fortification (wood-loam walls, moats, entrance gates, tower-like objects, palisades, stone walls, etc.).

b) production objects
- extent of the production centre and its position with respect to the settlement,
- location of the systems of pottery kilns, metallurgical smelt-houses and further production objects,
- positions of the individual objects,
- remains of mining activity (pits, shafts, galleries, etc.).

c) cult objects
- position of objects within the settlement,
- size, shape, orientation of objects,
- location of entrances,
- position of the sanctuary and/or further objects connected with it.

d) other objects (cellars, unvaulted cellars in loess, wells, etc.)
- run and location with respect to the settlement,
- size of objects.

e) architecture elements
- run of the masonry and its articulation.

B) Sepulchral Monuments

a) flat burial grounds (inhumation, cremation),
- position of the grave pit,
- properties of the grave, its arrangement (inhumation, cremation, inner structure) and its equipment (Fe-objects, etc.),

b) barrow fields (inhumation, cremation),
- grave character under the barrow and its equipment.

c) masonry tombs (stone, brick),
- position, size and groundplan arrangement of the object.

Relative complexity of geological relations of the Quaternary cover and the weathered material mantle (great lithofacial variability) and the resulting conspicuous changes in physical parameters of the environment studied, many times the complexity of hydrogeological and/or technical conditions (industrial interference, big recent iron and other objects near the areas studied, etc.) evoke the essential necessity of a broader choice not only of the suitable complex of geophysical methods, but also the processing and interpretation of the measured data using computer technology. In view of these facts and with respect to the overall character of works for the above types of tasks it is necessary to judge archaeogeophysical prospection as an activity of prevailingly investigation or semiinvestigation character oriented on a close co-operation with other specializations of both social and natural historical, technical and other branches of science.

The second block is oriented above all on the development and practical applications of programs for processing the measured data from magnetometry and the assertion/assesment of geoelectric methods on PC for solving different partial archaeological tasks. These ways of evaluation and interpretation - qualitative and quantitative - of the field data are described as being verified by practice, thus yielding a suitable basis for ranging and situating the excavation works. In this activity it was above all the monitoring of the properties of the rock environment in which the object studied of anthropogenic origin is found, its characteristic according to physical parameters and/or the creating of the corresponding physico-archaeological model of the structure studied or of a set of objects. In view of the fact that the third part of this paper/study/treatise summarizes the most important practical results from the field of archaeogeophysical prospection in the Czech Republic and abroad (Egypt, Germany) documenting the development of the application of geophysical methods for the individual types of the solved tasks, in the second paragraph only the main general methods are described for making programs and some practical specimens of processing.

The task/purpose of the publication is thus the analysis of the employed geophysical methods and the ways of processing (mathematical filtering, data transformation, etc.) with the submission of general regularities for their broader application in solving different tasks from the sphere of archaeological investigation in the field. Also further possibilities of application are suggested. Each partial task includes methodological examples of using the individual methods and the obtained (sometimes even less confirmative) results at our and foreign localities which are confronted with the data of subsequent excavation works.

From that point of view the paper contains a large amount of new knowledge derived from the analysis of the measured data, extending the information submitted in earlier contributions, and which has not been published so far. That concerns above all the investigation of fortification systems, architecture elements, burial grounds, etc.

## 3.1. Investigation in the Field

Prospection methods can, according to Wynn (1986: 533-537) generally be applied in

a) searching for an archaeological structure (site-exploration) - here aerial and infrared photography are used,
b) mapping inside the locality (intrasite mapping) - geophysical methods are mostly used only in this case.

The application of geophysical methods under suitable geological and technical conditions is one of the possibilities of achieving higher quality, speeding up and lowering the cost of archaeological investigation both in the Czech Republic and abroad (such as under conditions of the arid areas of Egypt, etc.). Classical excavation works represented above all by probing and/or shallow borings yield, besides a number of undoubted advantages, only point or a really limited information about the environment studied. Areal excavations are more time and finance consuming but the archaeologist gets often a larger number of data about the structure or object studied out of them, although on a smaller area. A rational complex of geophysical methods in a purposeful combination with all existing classical methods of investigation in the field makes a broader knowledge of the area studied as a whole possible (the extent of settlement on an open or a fortified settlement, the size of the burial ground, production centre, etc.), it extends and completes information about the key sectors of the locality before the archaeological excavation, by which the risk of the impairment of complicated stratigraphical situations in the beginning phase of the investigation is lowered. All these data are important for determining the efficiency and/or extent of the archaeological investigation in advance, and for determining the further progress of works in systematic investigations.

Geophysical methods used in solving the individual tasks (3.1.1.-3.1.3.) cannot be considered to be new, but only in recent years has their application reached a substantial extension in archaeological prospection. Geophysical works have become its inseparable part. Thus, since the last five years the representation of geophysics used in comprehensive archaeological investigations of the individual localities in the CR has increased more than three times.

The application of geophysics for the above actual objectives

within systematic long-term investigations and rescue activities was not only a methodological issue, but also a technical one linked up with the testing of accessible present or newly developed geophysical apparatuses and also the economic problem, given by the height of costs of the ground works. Before using geophysical methods for solving the required tasks it is therefore necessary to consider, whether the knowledge obtained will be proportional to the expended financial means, although it is known that geophysical information is relatively cheaper and in time substantially more quickly accessible than those of classical methods of archaeological investigation which can sometimes be of negative character, also in co-operation with geophysics (see e.g. Chapter 3.5.).

The choice and use of geophysical methods (magnetometry and geoelectricity) in the field investigation of the individual monuments of material culture is affected by generally valid conditions for their success which can be summarized into several main points:

a) a conspicuous differentiation of the environment studied according to physical parameters,
b) a suitable form, position and sufficient dimensions or thickness of the object studied (cultural layer) in a small depth below the surface of the field,
c) a relatively small shading effect of the overlying material,
d) a low level of interferences of natural or artificial origin.

From the above overview it is apparent that the effective application of these methods is not quite possible without a clear archaeological specification of objects of geophysical works, i.e. without the fundamental knowledge of the lithological characteristic of the environment, the character of the objects studied, their composition, depth of deposition, size, etc. From that there follows the necessity of a close and systematic co-operation of the archaeologist with the geophysicist from the elaboration of the project of field works up to the final evaluation of the whole activity.

The choice of the suitable geophysical method is conditioned both by the task of the field prospection and by its division into stages (Hašek - Měřínský 1991: 36). The application of geophysical disciplines in archaeological investigation of reconnaissance character (scale 1 : 200 - 1 : 1000) consists in the specification of the whole position, orientation and size of the structure (an open or fortified settlement, a Neolithic circular object - roundel, burial ground, etc. in the territory of interest about which a fundamental conception has already been obtained, e.g. by aerial photography, surface investigation, collection of pottery and further artifacts, phosphate analysis, etc. The extent of verifying sounding works is, as a rule, not large, and geophysical prospection thus complements and extends the fundamental knowledge about the spatial arrangement of the main anthropogenic structures.

In detailed research (scale 1 : 50 to 1 : 200) it is passed from broader archaeological issues to specifying the course and position of the individual members of the mapped structure, such as the following of entrance gates, a detailed structure of the body of the defence line - vallum in the fortification systems, the division of the habitation and further elements of settlement, the orientation of grave pits, location of the individual furnaces, galleries, shafts, tombs, cavities, masonry relics, etc. Field measurements are completed by both laboratory analysis of samples of rocks and in situ. The archaeological characteristics of the objects of interest of a rather descriptive character is extended by quantitative data. Geophysical measurements trench more broadly on the territory of the structure studied, completing mostly local and fragmentary information from classical excavation works to the whole space of interest.

The effect brought along by the purposeful combination of prospection methods is above all conditioned by the fundamental sticking to the stages of the works and a consistent application of the correct link-up of the individual investigation methods.

Magnetometry and geoelectric measurements applied in the CR in archaeogeophysical prospection can be divided into two groups:
> a) methods used in the investigation to the broadest possible extent (up to the mid-1980s) - proton magnetometry, DC geoelectric methods,
> b) methods utilized in field works since 1985 up to the present - electromagnetic methods (dipole inductive profiling, ground georadar, induction finders and/or the method of very long waves) and magnetometry - measuring of vertical gradients.

Note: Besides these methods applied for the solution of the required tasks also further ones are applied to limited extent, such as the shallow refraction seismism, microgravimetry, thermometry and gamma spectrometry (Hašek - Měřínský et al. 1991, 45-50) which, however, did not experience a broad distribution (see Chapter 3.1.3.).

The methods of field geophysical works are usually determined according to the principle to continue from the quickest and methods (such as magnetometry, DEMP and others) to methods relatively more complicated and costly (ground radar, vertical electric sounding-VES etc.). They must be oriented mainly on such a resolution as is sufficient for the spatial location of objects of interest of the given dimensions and depth. The resolution includes the density of the geodetic points etc. The optimum network of measurements is that which guarantees with certain probability the given tasks at a relatively low loss of information. When selecting the optimum distance between the profiles and the measuring points, the basis is the solved archaeological task. As reliably determined is such an anomaly as appears at a minimum of three parallel profiles in three measured points, so that the network of measurement is determined by the expected dimensions of anomalies along the area. By anomaly are understood such values of the studied field that exceed three times the mean quadratic error of measurement (Nikitin 1986, 94-96). From among programs elaborated by Varchomeev (in Tarchov et al. 1982) it followed that the probability $P_{3,3}$ >0.95 of the object found of the dimensions 3 x 3 m is, with the measurement step 1 x 1 m, at distances 2 x 2 m, $P_{2,2}$ only 0.6. In small archaeological objects - the individual furnaces, graves, storage vessels, palisade grooves etc. the density of profiles and measurement points must be higher. Thus, for a grave pit of dimensions 2

x 1 m $P_{2,1}$ is greater than 0.95 only with the network of measurement of 0.5 x 0.5 m. For bigger archaeological structures, such as fortifications, loam pits etc. the measurement network up to 2 x 2 m is sufficient from the above calculations.

### 3.1.1. Magnetometric Method

Magnetometry has become, due to its speed and high efficacy one of the most frequently employed methods in the reconnaissance and detailed archaeological investigation. The method of magnetometry was improved in 1974, after the introduction of proton magnetometers into geophysical practice fully replaced earlier, less exact and slower magnetic (torsion, edge) balances.

In its application it is above all the location of different anthropogenic objects differing by their magnetic properties from the surrounding Quaternary or earlier environment (see Chapter 3.4.), which permits their search by the above prospection method. The principles and use of magnetometry for the purposes of archaeology have been discussed in a number of publications, such as Aitken (1958, 1959, 1961); Breiner (1973); Frantov-Pinkevich (1973); Linington (1969: 131-138; 1970: 169-194); Logachev-Zakharov (1979); Scollar (1965: 21-92); Hašek - Ludikovský (1977a: 108-115); Hašek - Měřínský(1991: 37-40); Marek - Pleslová (1977: 55-59); Mareš et al. (1979: 147) and others.

Sources of magnetic anomalies are above all
> a) fireplaces, furnaces, burnt clays, fire layers, potsherd depositions, Fe-objects etc., i.e. monuments/relics whose magnetization has formed due to the magnetic field under conditions of considerable temperature changes. It is thermoremanent magnetization due to a strong heating of loams and clays with a content of magnetites to a relatively high temperature and cooling in the Earth's magnetic field,
> b) recesses secondarily filled with dark (fossile) loams with organic remains, ash filling, daub, pieces of coal, slag material, etc. (ditches, settlement and habitation objects, grave pits, loam pits, etc.),
> c) masonry elements of walls of magnetically anomalous rocks (granodiorite, diorite, gabbro, basalt, adobe bricks, etc.),

For the field works are used various proton magnetometers based on the measurement of the mean frequency of proton precession in the magnetic field which is proportional to the size of the operating magnetic field. Up to the mid-1980s only the size of the total vector of the Earth's magnetic field was measured.

The sensitivity of the instrument was ±1 nT. The measurements (2 to 5 readings/point) were performed in a regular network of profiles and points whose total density was dependent on the character of the task solved. In most cases the basic unit of dimensions 50 x 50 m was sufficient which is measured in a square network of 2 x 2 m, 1 x 1 m or 0.5 x 0.5 m, sometimes a rectangular network/grid of 2 x 1 m.

In parallel to the above measurement, at a suitable point, outside the impaired territory, there was a registration of short-time, above all daily variations, i.e. time changes in the magnetic field which, under normal conditions are the consequence of the magnetic effect of the system of electric currents in the ionosphere. Their measurement was carried out at regular intervals. The step of the registration, in dependence on the size of the short-period interferences (such as pulsation, BAY and industrial currents, etc.) varied mostly within the interval of 2 to 60 s. The measurement was performed above all by means of proton magnetometers G-816 of the firm Geometrics (USA), MP-2 of the firm Scintrex (Canada), PM-1 and PM-2 (products of Geofyzika, state enterprise, Brno). The height of the magnetometer probe varied on the average up to 0.6 m. As variation station there were mostly magnetometers PM-2 or PM-1. The reading of the data measured from the last two types of the above instruments was done automatically to a semiconductor logger for the subsequent evaluation on PC.

In the 1990s two interconnected magnetometers PM-2 with two probes located on one rod began to be used, and since 1994 a gradiometer GPM-1 for the direct measurement of vertical or horizontal gradients of the anomalous field. The distance of probes in both cases is changeable, mostly the distance of sensors of 100 cm is used (60 cm and 160 cm). In simultaneous measurement at two heights above the ground it can be assumed that the level of interference (electromagnetic, sign of variations, etc.) will be practically the same in the two probes and it will be subtracted from each other. The resolution of the instrument is 0.1 nT/m. Note: Abroad, e.g. in Austria etc. also caesium magnetometers are used in field prospection. They work with the accuracy of 0.01 nT/m. With respect to a complicated design of securing the measurement probes, this apparatus can be used mostly in a bare field without vegetation and in places of low industrial interference.

Field works are carried out in the same network as in the current profiling method, at one point being carried out 5-10 repeated measurements according to the interference.

The advantage of this system of measurement is the fact that
- a) it can be carried out even under difficult natural conditions, in places strongly affected by industrial and other activity, where it would not be possible to use current magnetic methods,
- b) it divides the complicated anomalies $\Delta T$ into individual parts,
- c) it automatically excludes the regional gradient,
- d) it marks better shallow sources of magnetic anomalies,
- e) it removes variations of the field and suppresses interferences.

A drawback is, however, the necessity of a higher number of repeated measurements at one point and a greater dependence on the orientation of the magnetometer probes.

For measuring the apparent magnetic susceptibility ($æ_{ad}$), e.g. on the walls of the individual exposed objects, rock outcrops etc., the kappameter KT-5 is used (product of Geogyzika Co., Brno) with a numerical output (calibrated in SI units, sensitivity $1.10^{-5}$ j.SI) which allows to draw the last 12

measured values from the memory of the apparatus.

### 3.1.2. Geoelectric Methods

Some of the methods of geoelectric investigation are used to a great extent both in the CR (Bárta 1973: 6-7; Bárta et al. 1980: 9-13; Bárta-Marek-Pleslová 1987: 10-20; Bílý 1983: 133-138; Gruntorád-Karous 1972; Hašek - Měřínský1991: 40-45; Mareš et al. 1979: 247; Tirpák 1977: 120-122, Záhora 1989: 226-252, and others), and abroad (Clark 1986: 1404-1413; Frantov-Pinkevich 1973: 8-100; Frohlich-Lancaster 1986: 1404-1425; Lerici 1955: 1-2, 1960; Tabbagh 1974: 350-437; Vaughan 1986: 594-604; Weymouth 1986: 538-552; Wynn 1986: 533-537, etc.) in field prospection of objects of anthropogenic origin characterized by an increased differentiation of specific resistance and/or conductivity and relative electrical permittivity towards the surrounding environment (see Chapter 3.4.).

By means of geoelectric methods (DC and electromagnetic) the following tasks are usually solved
- a) mapping of fortification systems (defence line - vallums, walls, moats), remains of mining activity (galleries, shafts),
- b) following the foundation walls of different stone objects, elements of architecture, overall extent of the cultural layer of settlements, their individual parts,
- c) location of barrows, flat burial grounds, tombs, etc.

At present many modifications are known, differing by the excitation and measurement of the electric field. In further paragraphs of this chapter only those methods are described whose efficiency in archaeological investigation has been confirmed by practice and which can possibly be recommended for a broader application for solving different tasks of these issues.

#### 3.1.2.1. Direct Current Resistance Methods

*I. The Method of Resistance Profiling (RP)*

One of the fundamental geoelectric methods used in archaeogeophysical prospection since the late 1940s has been resistance profiling (Atkinson 1952). Its different modifications make it possible to find the values of the apparent specific resistance ($\rho_a$) in an inhomogeneous environment at a constant distance of the current (AB) and the measuring electrodes (MN). At the individual points the potential difference or voltage between the near measuring electrodes ($\Delta V$, mV) is read. The direct or alternating current of low frequency (I, mA) is introduced to the ground by means of current electrodes. Then for $\rho_a$ (ohmm) = K. $\Delta V/I$, where K is a constant of the arrangement (m) depending on the distance of the individual electrodes. The calculation is given in a number of publications, such as Gruntorád-Karous 1972; Mareš et al. 1979: 262-283, etc..

In measuring the chosen arrangements of electrodes mostly moves along the profile and/or the neighbouring profiles, preserving their mutual distance, i.e. also the constant of arrangement. Lateral changes are monitored in the environment below the surface and thus vertical interfaces are found, i.e. inhomogeneities of different specific resistances. Their differences can be explained either as a consequence of lithofacial changes in the environment studied or by the position of the anthropogenic object with anomalous (increased or decreased) specific resistance in the soil or rock layer (massif). In differential arrangements zero potential difference is measured in the homogeneous environment or the zero potential difference (constant arrangement = infinity). They are used for the location of conspicuous nonconductors and conductors in the homogeneous environment. Archaeological structures (objects) are located by the so-called "pure anomaly".

The field profile measurements are usually carried out at different electrode configurations, both potential and gradient ones and depth interventions. The most frequently applied arrangements are given in Fig. 1.

The measurement density in the individual electrode configurations depends on the character of the task solved, the size and distance of deposition of the object of interest. Purposeful is the regular network of profiles, also perpendicular to each other, oriented as far as possible perpendicular to the assumed archaeological structures. The step of measurement is recommended to be 1 - 2 m, the distance between the profiles in the reconnaissance measurement being 10 - 20 m, in detail works 2 - 5 m.

The starting data for the processing of field data are graphs of the measured or filtered measured magnitudes ($\rho_a$, $\Delta V/I$), maps of isolines, etc.

*II. Vertical Electrical Sounding (VES)*

The method of VES permits us to quantitatively determine the thicknesses of differently conductive layers and the regularities of changes in specific resistances of the rock environment in the vertical direction. In archaeogeophysical prospection the Schlumberger arrangement is used with $MN/2 = 0.2, 1, 5$ m, the distance of current electrodes and thus also the depth range of the current is increased for $r = AB/2$ from 1 to 132 m at the step $r_{i+1} / r_1 = 1.2$. The maximum distance AB depends on the characteristic of the geoelectric section and it is, as a rule, 5 and more times greater than the depth of the studied interface. The distances of the individual points of the VES on the profiles vary on the average from 5 to 20 m.

The instruments used in the Czech Republic for resistance measurements (RP, VES) are above all the apparatus GESKA-76 and the unit complex GEOVYS (MINI-II, GEVY 100). MINI-II is a millivoltmeter and a compensator SP. Thanks to efficient filtering of the interfering voltage the instrument works reliably even at greater distances of the current electrodes in regions with a high level of interference - errant, industrial currents etc. GEVY 100 is one of the basic units of the transmitter system. The highest output of the transmitter is 100 W, the voltage 620 V, the current 316 mA. The transmission output, voltage and current is wireless controlled stepwise and continually and simply remotely switched off. The employed electrodes are copper (MN) and iron (AB) and rod shaped (diameter about 0.8 cm).

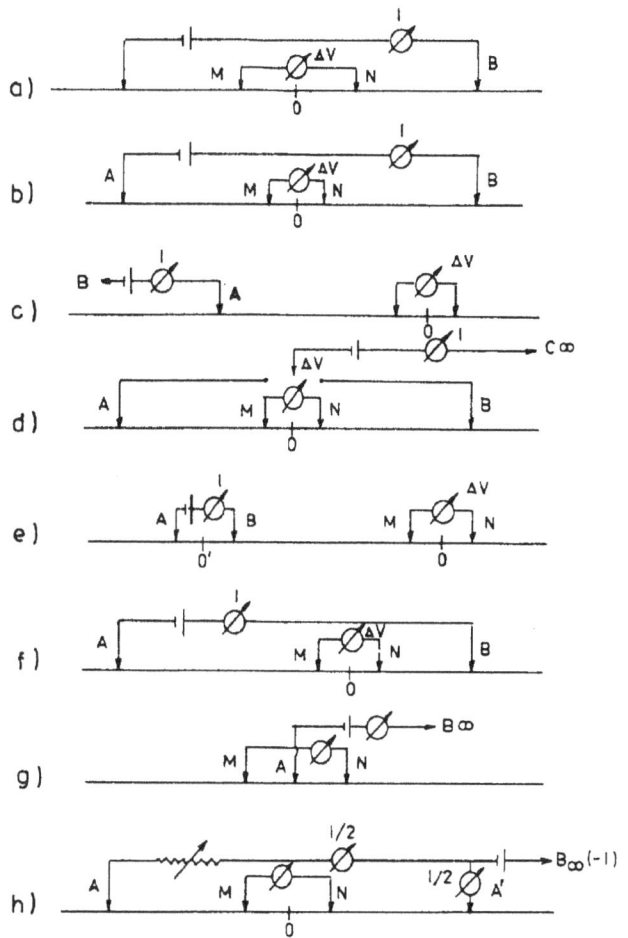

**Fig. 1.** *Electrode arrangement recommended for resistance measurements in archaeogeophysical prospection of the CR a) the Wenner profiling (AMNB), b) the Schlumberger profiling (AMNB), c) three-electrode gradient profiling (AMN), d) combined profiling (AMN, MNB), e) dipole axial profiling (ABMN), f) central gradient method (AMNB), g) differential potential profiling (MAN), h) differential gradient profiling (AMNA), AB- current electrodes, MN - measuring electrodes, O - point of record*

The starting material for the interpretation of VES are the curves $\rho_a$ (ohmm) = f (AB/2) (m). The result of qualitative processing which gives an idea of geoelectric conditions of the territory investigated are the apparent isometric sections, when the specific values of $\rho_a$ are plotted under the terrain relief to the depths $h = r/2 = AB/4$. From the quantitative interpretation (computer or graphic evaluation) physico-geologic profiles with the given thickness and resistances of the individual layers (in archaeogeophysical prospection it is prevailingly a three-layer environment), together with further information about the structure studied, or maps of isohypses of the relief (and/or the thickness) of the studied horizon are drawn.

### 3.1.2.2. Electromagnetic Methods

In the last decade, besides magnetometry or geoelectric DC methods, also electromagnetic methods have been used on a larger scale in the CR in solving some partial tasks in archaeological prospection which have a broader application above all in the regions strongly affected by industrial and other activity.

They are above all
   a) the method of dipole electromagnetic profiling (DEMP) which started to be used for these purposes in this country after obtaining conductometers EM-31 and EM-38 (Hašek - Měřínský 1987: 102-140).
   b) the resistance modification of the very long wave method (VLW-R), originally intended for engineering-geological mapping and ore prospection (Bláha-Chyba 1978; Karous 1982: 77-79, etc.), which can be applied for the solution of different tasks of a more regional character, such as finding the extent of the studied archaeological structure in great depths, the character of the rock massif, etc. (Hašek et al. 1982a, 1988).
   c) radiolocation method (GPR) used mainly for the location of different inhomogeneities near the surface, such as cavities, building foundations, etc. (Bílý 1983: 133-138; Hašek et al. 1985; Hašek - Unger-Záhora 1997).

*III. The Method of Dipole Electromagnetic Profiling*

Electromagnetic profiling with a small dipole source (DEMP or DIP) belongs to the low frequency methods with a mobile source. The possibility of direct and contactless finding of the resistance properties of the environment follows from the dependence of the measured components of the electromagnetic field on the apparent specific resistance of the equivalent conductive semispace.

Using DEMP in archaeogeophysical prospection has been described in foreign literature since the early 1980s, such as Banning et al. (1980), Bevan (1983: 47-54), Frohlich-Ortner (1982: 249-267), etc. The method is employed mainly in finding different nonconductive and conductive objects of anthropogenic origin, similarly as RP.

Due to the possibility of continual and point measurement, high productivity and the demand for lower manpower necessary for carrying out field works it replaces partly, and in some cases completely, the method OP, completes VES (measuring at ZZ and YY polarization, at different heights, etc.) and the soil radar.

The time variable magnetic field evoked by the AC, induces in the transmitting coil in the conductive rock environment very weak rotational currents which form a secondary magnetic field Hs phase shifted by $\pi/2$ as against the primary field registered in the receiving coil together with the primary field Hp. This secondary magnetic field is generally a very complicated function of the distance between the dipoles r, the employed frequency f and the conductivity of the environment σ.

$$(Hs/Hp)_v = 2\{(Kr)^2\{9-[9Kr+4(Kr)^2+(Kr)^3]e^{-Kr}\}\} \quad (1a,b)$$

$$(Hs/Hp)_H = 2\{1-3/(Kr)^2+[3+3Kr+(Kr)^2]e^{-Kr}/(Kr)^2\}$$

$$\sigma_a = 4 / j\,\omega\,\mu_0\,r^2\,Hs/Hp \quad (2)$$

where $K = - (j\,\mu_0\,\sigma)^{1/2}$ ............... wave number
$j = (-1)^{1/2}$

$w = 2\pi f$, $\mu_0$ ..................... vacuum permeability
r ....................................... distance between dipoles
$\sigma_a$ ................................. apparent conductivity

The unit of apparent conductivity is the millisiemens (mS.m$^{-1}$).

For the transfer of $\sigma_a$ to the apparent specific resistance the folowing formula is used

$$\rho_a (\text{ohmm}) = 1000/\sigma_a \quad (3)$$

In fieldwork conductometers are used for archaeological purposes whose rigid measuring system is constituted by a transmitting and a receiving magnetic dipole. Their distance limits the depth range of the apparatus, designed for small depths (max to several metres). The vertical components of the field are measured, monotonously dependent on the specific resistance of the environment. The measured values can be transformed to data of apparent conductivities or resistances (see formulas 2 and 3).

Conductometers - the digital KD-1 (CS product) and/or EM-31 made by Geonics Limited Canada have firmly connected dipoles distant 3.66 m and they work at the frequency of 9.8 KHz. EM-38 of Geonics has parameters f = 13.2 KHz, r = 1 m.

The maximum depth range stated by the producer in EM-31 (KD-1) is 3 - 5 m (polarization ZZ) and about 2-3 m (polarization YY), in EM-38 for the horizontal component 0.75 m, for the vertical one 1.5 m. The step of measurement at the profiles is 1 to 2 m.

*IV. The Resistance Version of the Very Long Wave Method*

This belongs among the passive electromagnetic methods - the transmitter is at a great distance from the receiver. For the purpose of solving the issues of inhomogeneities near the surface this method was used for the first time in this country in 1981 for finding the positions of galleries of old mining activity in the region of the North Bohemian Lignite Area (Hašek et al. 1982a). In archaeogeophysical prospection it is employed despite its rate and productivity for the time being only to a limited extent. In principle it is based on the peculiarities of the propagation of radio waves which, near the surface, is affected by the geological structure of the rock environment. By measuring some parameters of electromagnetic fields of transmitters it is possible to reversely determine the electrical properties of rocks at the position of the receiver. Due to the distance of the receiver from the transmitter (the present electromagnetic fields of strong navigation radiostations are utilized in the VLW band of f = 15.1 to 22.16 KHz) it can be assumed that the electromagnetic field has the character of a plane wave. If the measurement is carried out above the homogeneous semispace, the vertical magnetic component can be assumed to be negligible. It will appear only at the presence of conductivity inhomogeneities in which the secondary field is induced.

The electromagnetic field is monochromatic; it has a constant approximately linear polarization. The impedance Z of the plane wave, polarized along the x axis and propagating in the direction of the positive z axis, can be expressed in the form

$$Z = Ex/Hy = (-i\omega\mu\rho_{VDV})^{1/2} \qquad (4)$$

By adapting in expression (4) a formula for specific resistance is obtained

$$\rho_{VLWV} = iZ^2/\omega\mu = 1/\omega\mu \, (Ex/Hy)^2 \qquad (5)$$

where ω.      circular frequency
μ      permeability
E      intensity of the electric field
H      intensity of the magnetic field

The formula (5) calculates resistance when applied for measuring over a homogeneous semispace.

The complementing information about the geoelectric parameters of the environment can be found from the phase shift between the electric field and the magnetic one $\varphi$ ($^O$), thus

$$\varphi(^O) = \varphi_{Ex} - \varphi_{Hy} = argZ \qquad (6)$$

The VLW-R method can be applied in searching for steep and obliquely slanting nonconductors and conductors in the system of two different arrangements - H and E (Bláha - Chyba 1978). In archaeogeophysical prospection particularly H-polarization is utilized, when the electric dipole (of the measuring electrode) is oriented along the profile, i.e. perpendicular to the structure. The component of the fields Ex and Hy is measured. The source situation in this case is the transmitter in the direction of the profiles. The magnetic component Hy is parallel to the direction of the inhomogeneity. This modification is suitable for the location of elongated nonconducting structures and subhorizontal conductors. At E - polarization, on the contrary, the electric dipole is situated perpendicular to the profile, i.e. parallel to the linear archaeological structures. Components Ey and Hx are measured. This system of measurement is more sensitive for finding conductors and the values of anomalies are less distorted by horizontal near-surface inhomogeneities, such as the cover, etc. For field works the digital apparatus VDV-1 is used (product of Geofyzika Co. Brno). The step of measurement is usually 2 to 5 m on the profiles.

*V. Geophysical Radiolocation Method*

This method was used in archaeogeophysical prospection in the former Czechoslovakia for the first time in 1982 (Bílý 1983: 133-138). Abroad were these issues dealt with by Kenyo (1977: 48-55); Moffatt (1974); Money (1974: 213-232); Vaughan (1986: 594-602); Weymouth (1986: 538-552); Wynn (1986: 533-537) and others. It is used above all for finding tombs, cavities and other near-surface inhomogeneities, above all in the built-up area.

The transmitter of electromagnetic waves of the frequency of 50 - 400 MHz and at the same time also the receiver of the echo is the movable aerial placed immediately above the ground. The apparatus receives the echo and processes it at the suitable recording device in order to make it possible to determine the time of arrival of the individual waves from the time of transmission of the electromagnetic pulse after plotting the reflected waves. In field works the aerial is shifted above the studied rock environment along a certain stepping distance, or it is moved along the profile. The result is the time section of the measured profile (an analogy of the seismic record) obtained already in the course of the measurement. It can be preliminarily evaluated directly at the locality.

The practical use of the GPR is based on the found different relative permittivities and apparent specific resistances of the individual objects studied. A disadvantage of the method is its limited depth range (about 4-6 m). The increase of the middle of the frequency band improves the resolution of inhomogeneities, but it reduces the depth of penetration.

The apparatus used in the Czech Republic in archaeogeophysical prospection consist in principle of two main parts - the control unit and the aerial. The auxiliary parts are formed by the feeding unit, the plotter, a tape recorder with accessories, etc. They are the Canadian PULSE EKKO and RAMAC/GPR of Swedish production. The field measurements are made on individual profiles, specified according to the results of other applied methods, such as DEMP etc. (Hašek - Unger - Záhora 1997). Note: For finding small recent and archaeological Fe-objects in the near-surface layer (a disturbing element for geomagnetic measurements) and further non-ferrous metals various detectors of iron and non-iron metals are used, such as MSG-4 of the firm Severin (Germany) which works in the balance system (inductive balance with the depth range of max. 40-60 cm) or TM-91 DISCOVERER of the firm Geofyzika joint stock Co., WEIT SIEGEL SPECTRUM.

### 3.1.3. Other Methods Applied

Besides the methods described above, which are used to a large extent for solving different tasks of archaeological prospection and investigation, also further geophysical disciplines are used in the CR, but due to their difficulty in measuring, processing and due to higher costs they have not spread. They are above all applications of gravimetric, seismic, radiometric and geothermic methods (Hašek - Měřínský 1991: 45-50). For the sake of completeness an overview of methods of geophysical works, the processing of measured data and the possibilities of their practical application in archaeogeophysical prospection is given here.

Gravimetric methods serve for delimiting objects of anthropogenic origin with a mass different from that of the surrounding environment. They are used particularly in finding positions of cavities of different origin, such as tombs, galleries, etc. At topographically measured points relative changes in gravitational acceleration is measured with gravimeters with respect to the data at cardinal gravimetric points. The gravity changes in the homogeneous environment with the height above sea level of the measured point and with its geographical latitude. In an inhomogeneous environment the changes in gravitation are further affected by density inhomogeneities occurring in the neighbourhood of the measured point.

For field works gravimeters with a quartz system are used, which allows us to measure the gravity with the accuracy of

± 0.15 μms$^{-2}$. Bouquer's gravitational anomalies due to density inhomogeneities of the archaeological objects searched for achieve, according to modelling the values of 0.1 to 2.0 μms$^{-2}$ (Hašek - Měřínský1991: 45-46). The gravity points in the space of interest are laid out on the profile with the step of 0.5 to 3 m.

The processing of the measured data is at present made on minicomputers or PCs according to the compiled system of programs (Blížkovský et al. 1976; Odstrčil 1985, etc.).

The shallow refraction-seismic method (sometimes also shallow refraction) applied experimentally in archaeology, above all in searching for fortification systems and near-surface cavities is in essence coincident with the methods of measuring for the purposes of IG. In most cases it is operated with small depths of 2 to 5 m. The correlation method of broken waves is used. The methods of works start from the measuring at longitudinal and transversal profiles. From the analyses of the wave picture, the configuration of the hodochrones and the analysis of the rate of the registered waves it is possible to judge at the depth of the seismic interface (e.g. the filling of a moat, etc.), its morphology, state and composition of rocks near this limit, etc.

Finding the position and dimension of the archaeological object by the seismic method is possible only in cases where it is located at such depth that there is not yet a complete absorption of the signal. With respect to the present instrumental possibilities it can be judged that the depth of the foundations of the objects investigated should not exceed 3 - 5 m. It can be assumed that the rate of propagation of seismic waves in a Quaternary cover varies on the average from 800 to 1200 m.s$^{-1}$ and the dimensions of the objects are 2 x 2 m or 1 x 1 m. For those reasons the length of the wave must be of the order of 50 cm and f = v/λ = 1.6 - 2.4 KHz, i.e. that archaeological seismism requires higher frequencies in comparison with the purposes of IG.

In the method of shallow refraction in archaeology three modifications are used:
   a) continuous profiling, based on the measuring along the profile at a constant distance between the points of the blasting and geophones (the step of geophones 1 - 2 m, distances between the points of detonation 50 - 200 m),
   b) sounding at measuring at selected points, such as crossings of profiles in several directions (following the rate of anisotropy),
   c) fan-like determination of rate directions (circle diameter about 10 - 20 m, the distance of geophones along the circumference 1 - 2 m.

In the method of penetrating by seismic waves (location of cavities) the measuring is between boreholes and between the borehole and the surface.

For shallow seismic works special one-channel or multichannel apparatuses are used, most of them with analog registration. For solving the individual tasks it is possible to use the following instruments: ABEM TRIO/24 and 12 channels, 1050A of the firm OYO (24 channels) BIZON/1 and 6 channels. The excitation of the seismic energy is carried out by the stroke of a hammer on the support plate,

by the falling weight and the explosive located in shallow wells on the surface of the ground, etc.

By processing the seismograms hodochrones of direct and broken waves are obtained. From the hodochrones of direct waves the rate in the cover is obtained; hodochrones of broken waves yield data about the rates of rocks in their substrate. According to the values of rates it is possible to judge the character of the near-surface seismic interfaces, and/or it is possible to find the degree of the impairment of the massif. The main result of processing is the depth course of the interface between different lithologically discrepant types of rocks (Hašek et al.1982b, 1985; Hašek - Veselý - Woznica 1981: 83-126, etc.).

Radiometric methods have not found acceptance in archaeological prospection, besides some experimental works (such as Čepela 1989: 49-64). However, they are used on a large scale in the complex of methods for measurements in shallow boreholes. They can be used in mapping various kinds of rocks and objects made of building materials with higher radioactivity (Mareš et al. 1979: 148-150).

For field works on the surface gammaspectrometers are used , e.g. GR 410 of the firm Geometrics and GS 256 (product of Geofyzika Brno Co.).

Geothermic methods measuring the natural thermic field and its gradient, thermal conductivity and the thermic flux can be included among prospective methods of archaeological investigation. They can be used for locating above all near-surface cavities (cf. e.g. Hašek et al. 1981; Müller-Müllerová 1976; Uhlík 1968, etc.). For the field works it is possible to use thermometers GT-1 and GT-2 (products of Geofyzika Brno Co.).

## 3.2. Processing of Measured Data

For the evaluation of a large amount of field data, particularly from profile magnetometric and some geoelectric measurements (OP, DEMP, VLW-R) and/or for their interpretation or reinterpretation, besides the application of the existing graphic ways of processing (Mareš et al. 1979, etc.) a number of programs for PC were made (Halíř - Hašek 1989: 193-205; Hašek et al. 1988a; Hašek – Měřínský 1989b; Hašek - Segeth-Vencálek 1990: 156-192) solving some tasks of the problems of:
   a) evaluation of the measured data and graphic output correction (profile curves of the measured magnitudes, maps of isolines and further displaying ways),
   b) qualitative interpretation (separation of regional and residual anomalies, calculation of correlation coefficients, etc.),
   c) quantitative interpretation (modelling, methods of the best coherence of the measured anomaly with the physical field of the calculated model, comparison with the theoretical curves or by means of nonlinear optimization algorithms, deconvolution, etc.).

The above programs (Table 1) were compiled for all types of computers used by various geophysical institutions in the CR.

As proved by Nikitin (1986), there exist in essence two main approaches to the processing and interpretation of data from the results of field measurements of different geophysical methods - the determined method and the statistical one.

The methods of the determined way follow from the assumption of direct connection between the values of the measured physical field and the disturbing bodies (near-surface inhomogeneities, atmospheric and other effects, etc.). These analytical methods of an only solution of the task are based on the theory of potential (magnetometry, DC resistance methods), Maxwell's equations (electromagnetic methods) etc. Their application makes it possible under favourable conditions (intense anomaly of the measured field, apparent differentiation of physical parameters, etc.) not only to find anomalous bodies, but also to evaluate their parameters.

Methods of the statistical approach to geophysical data have been recently extensively developed. It is connected with their characteristic consisting in the fact that the field data in the individual points must necessarily be considered to be accidental magnitudes. Accidental is also the position of different objects, points, areas of investigation, etc. Particularly the noise due to errors in the measurement, small inhomogeneities in the near-surface layer, variations of fields, etc. creates the accidental character of the physical field. In its evaluation one can apply the theory of probability and other mathematical and statistical methods. The result of the processing are statistics, their estimates etc. Using some statistical methods permits us to separate the useful information from the data measured also in those cases when bodies of anthropogenic origin evoke anomalies whose intensity is comparable to the size of interferences - noise.

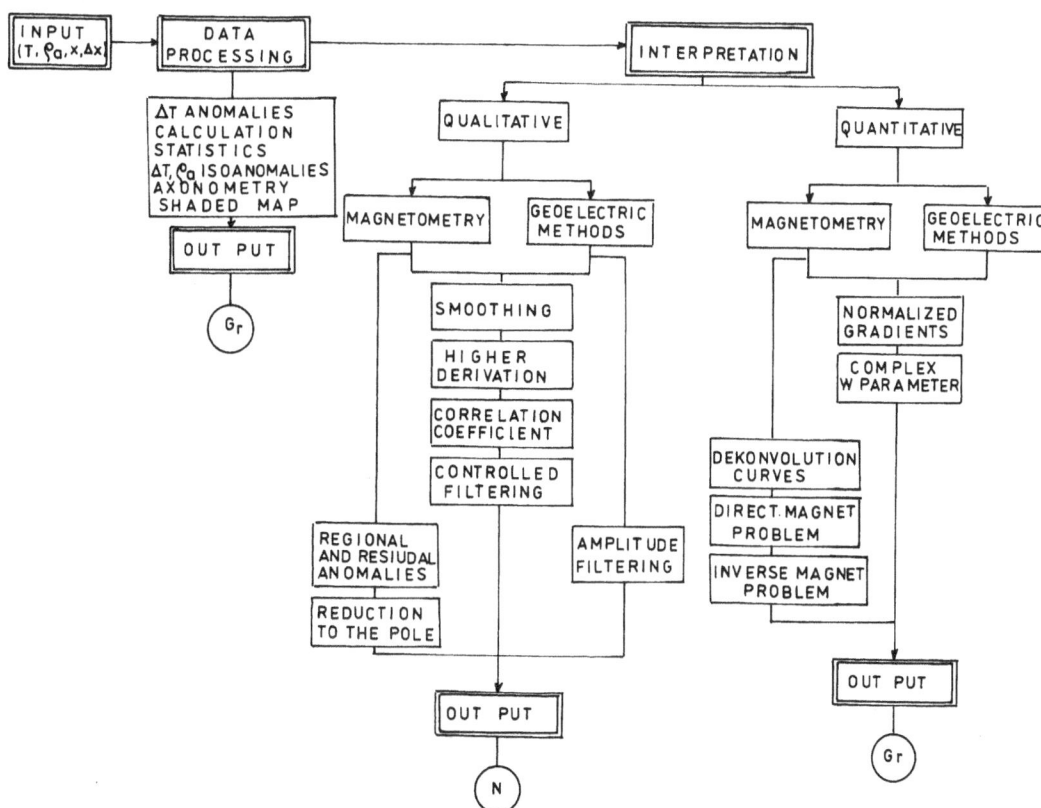

**Table 1.:** The general and simplified diagram for the processing and interpretation of profile measurements (magnetometric and geo-electric).

The measured data (input data) from magnetometry or geoelectricity are stored in the memory of the PC which makes the primary evaluation from the memory media of digital apparatus (such as gradiometer GPM-1 or by means of keyboards of field recorders (conductometer KD-1).

Their subsequent processing is done by means of partial programs which can secure e.g. corrections of erroneously stated values, the calculation of the normal relative field, reading of variations, etc.

The programs for graphical representation of field data or derived fields allow the drawing of profile curves, isolines, and their perspective representation. The plotting of the maps of isolines ($\Delta T$, $\partial T/\partial Z$, $\rho_a$, $\sigma_{DEMP}$, etc.) is divided into the interpolation of data by bicubic splins into a regular network, the determination of the position of points at which the field acquires the set values (data of the represented isolines) and their drawing. Since in the archaeogeophysical prospection the measurement is made prevailingly in the network 1 x 1 m In Fig. 2 the map of grad. $\partial T/\partial Z \equiv (T_z)$ of the locality or 2 x Troskotovice (district Znojmo) is given, which is an example of the utilization of this program.

The colour filling of areas between the isanomales $T_z$ (Fig. 3), the areal representation of the measured data (Hašek -

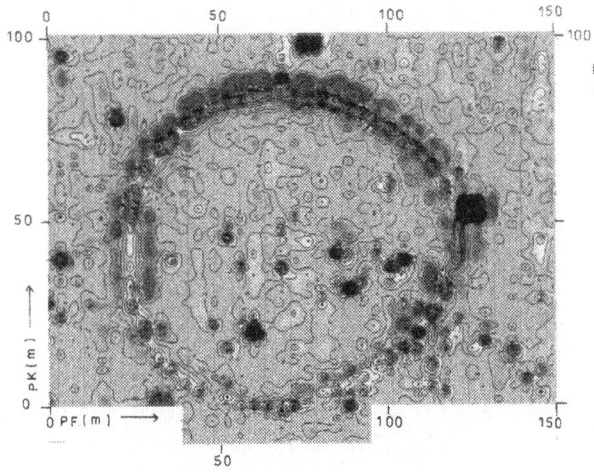

**Fig. 2.** *Map of grad ∂T/∂Z from the locality of Troskotovice, district Znojmo*

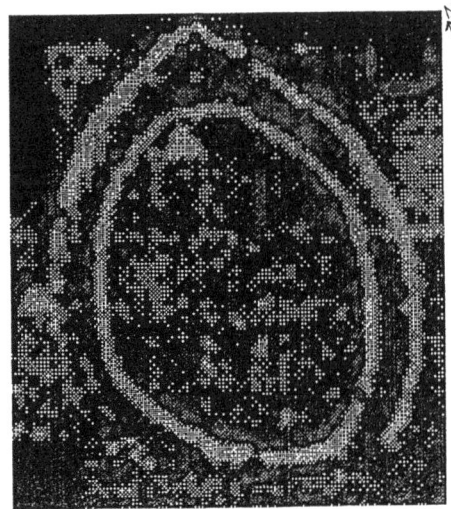

A )

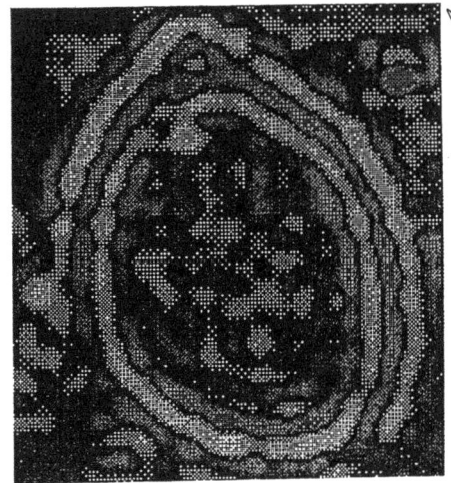

B )

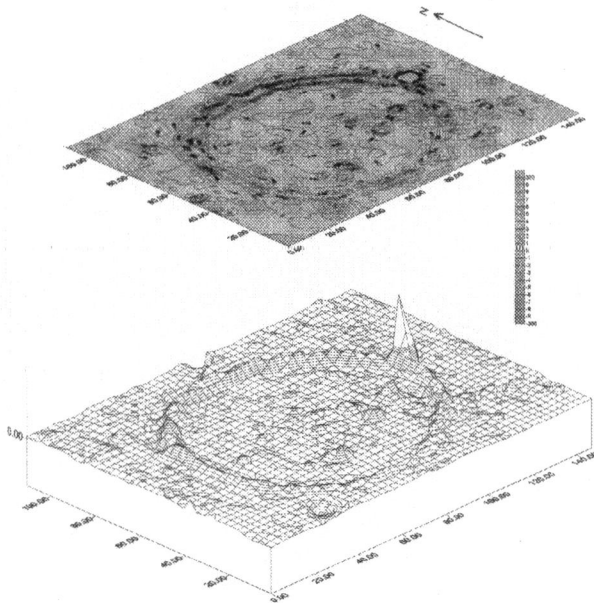

**Fig. 3.** *Colour processing of the map of ∂T/∂Z from Fig 2*

**Fig. 5.** *The application of the Fourier transformation and frequency filtering from Fig. 4*
*a) map of basic data*
*b) map of filtered data*

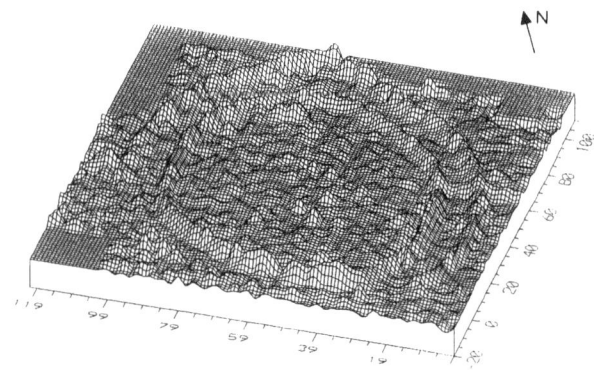

**Fig. 4.** *Areal representation of isanomales ΔT from the locality of Šumice, district Znojmo*

**Fig. 6.** *A black-and-white shadow map of ΔT from Fig. 4*

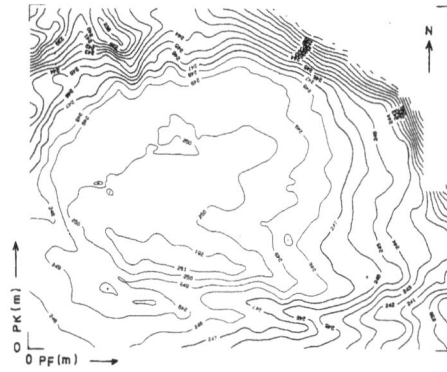

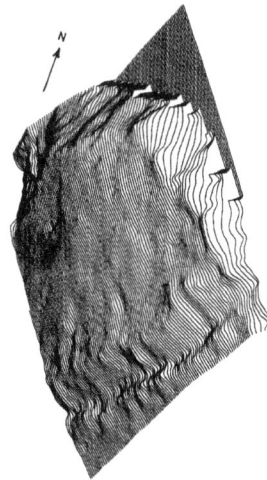

**Fig. 8.** *Geodetic processing of elevation data from the prehistoric hill fortification of the settlement of Kokory, district Přerov*

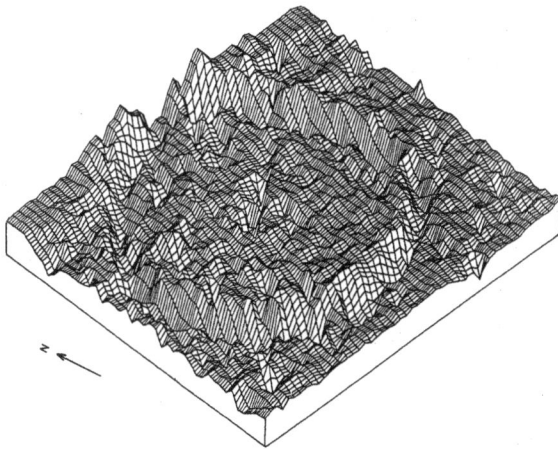

**Fig. 7.** *An example of the spatial representation of anomalies ΔT from the locality of Vedrovice, district Znojmo*

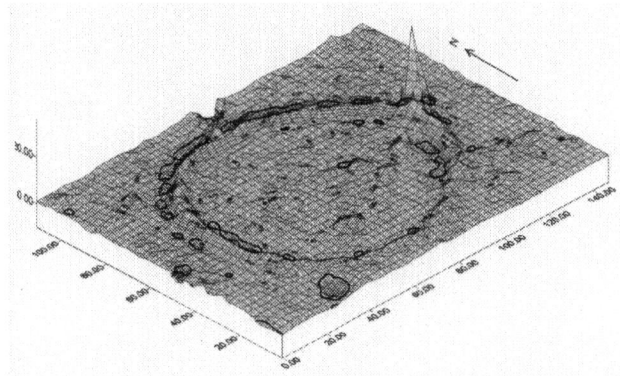

**Fig. 9.** *Axonometric presentation of neolithic circular structure, locality Trostkotovice, district of Znojmo*

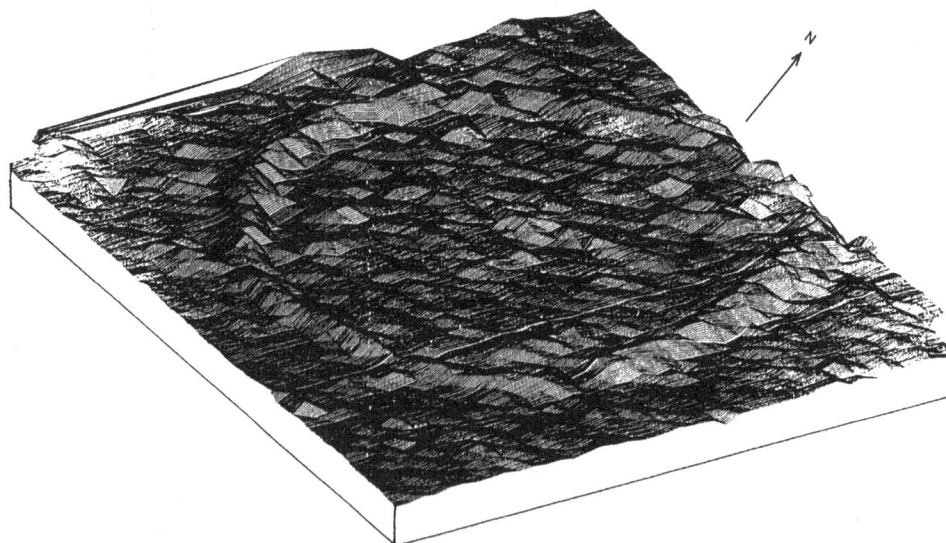

**Fig. 10.** *The map of ΔT profiles from the locality of Vedrovice, district Znojmo, projected to the spatial system of sections*

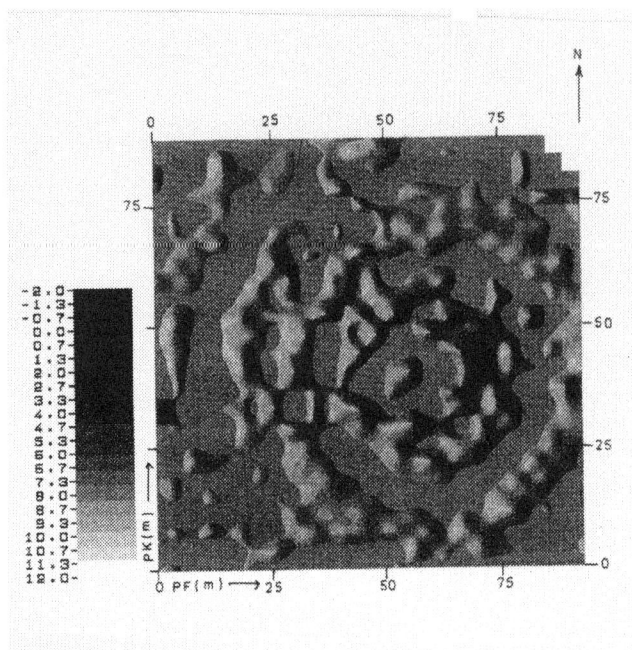

**Fig. 11.** *Relief illumination ΔT with an artificial source at the locality of Šumice, district Brno-country*

Tomešek 1995) (Fig. 4), and/or the application of the Fourier transformation and frequency filtering (Figs. 5 a,b) (Hašek - Petrová - Segeth 1994: 63-66). Further representation of anomalies T by a shadow map, compiled by means of bicubic splins is submitted in Fig. 6.

In some cases, particularly at archaeological localities with intense anomalies of the measured values it is reasonable, besides the maps of isolines, to process data also into the perspective view (Hašek - Horák-Obr 1979: 46-60; Hašek - Měřínský 1989: 103-149, etc.).

For the graphic representation of the function of two variables $Z = f(x,z)$, or a plane in the space, the program system G3 SRF was used which was developed in the

Institute of Physics and Mathematics, Academy of Sciences, Czech Republic (AS CR) in Brno. It allows to draw the systems of the selected sections of the surface with the planes $x = const.$, $z = const.$ and $z = const.$, and/or any of their combinations. These systems of lines are projected into the projection plane. Only the visible parts are drawn. The possibility of choice is from the following kinds of projection - parallel normal projection and central projection. The choice of the projection plane, and/or in the central projection also the centre of projection, different views of the surface studied can be achieved. If the field measurement is set only at node points of the network, as it is in the case of processed magnetometric data of the locality of Vedrovice (district Znojmo) (see Fig. 7), it is necessary to approximate the surface of interest. In the above case approximations for sections of the surface of splins of the 2nd order were used.

A similar program for representing the digital model of an archaeological locality or only an object was compiled at PÚDIS Prague for discrete points constituting an irregular triangular network. The input data are corrected as early as in the first regimes of the program. The applied methods of the axonometric representation and linear interpolations allow its construction from an irregular as well as regular network (Eisler-Pejša- Preuss 1988: 109-132). That contributes to the utilization of the method not only for processing the outputs based on geodetic measurement of the state of the Earth's surface before the investigation (Fig. 8) and the archaeological objects found in the course of excavation works (Fig. 9), but it can also be applied for evaluating the measured data from different geophysical disciplines (see Fig. 10).

The digital model in Fig. 9 expresses the main parts of a church structure (preliminary processing) in the state of preservation found by an excavation on the basis of results of geophysical works (see Fig. 89).

Also the results of magnetometric measurement in the space of the Neolithic circular area of the period of the Moravian painted pottery culture at Vedrovice (district Znojmo) in the form of profiles projected into the spatial system of sections (Fig. 10) (for comparison see also Fig. 7) have proved the expediency of this method for the solution of the above task.

Another method applied in this phase of processing consists of the representation of the measured data in the form of relief with artificial illumination. The surface of the relief that is turned towards the source are light, the averted ones are dark (Kovalik-Glen 1987: 875-884). A practical example of the evaluation of magnetometric data according to the above method at the locality of Šumice is given in Fig. 11.

## 3.3. Interpretation of the Measurement Results

### 3.3.1. Qualitative Interpretation

The qualitative interpretation is based on the division of the measured integrated field (e.g. geomagnetic) into local (archaeological or shallow disturbing objects) and regional anomalies (deeper geological sources) and into broken anomalies due to noise (inaccuracy of measurement, artificial sources, errant currents, etc.).

In geoelectric methods it is above all the finding of the character of the structures of anthropogenic origin, positions of nonconductive and conductive bodies and the possibility of their location by the selected arrangement of electrodes, the following of the depth range of the system used, the calculation of coefficients of mutual correlation, controlled filtering, etc.

#### 3.3.1.1. Magnetometric Method

A number of methods (Mareš et al. 1979) and programs (Hašek et al. 1988: 195-200) serve for the separation of useful anomalies. They use in essence the same principles as gravimetry. Either measurement on a high pole is done, or graphical differentiation, or from formal mathematical methods convolution in an optional sliding window according to (Griffin 1949), methods of higher differentiations (Lyubimov-Lyubimov 1983: 130-136), conversion of fields to a higher or lower level, wave length filtering methods, approximation of the anomalous field by the functional relation, bicubic splins (Pretlová 1976: 168-177) etc. The above methods of this mathematical processing create ring-like arranged fictive anomalies of opposite sign around residual anomalies, they lower their amplitude and reduce their wave length. This is due to the calculation method, the unsuitable dimension of the window, etc. (Hašek - Měřínský 1991: 55-58). Anomalies obtained by formal mathematical methods can only serve for the purpose of qualitative interpretation (Tomek 1975).

The generally accepted model of the geophysical field is the additive model - i.e. a model in which the results of the field f(x) measurement are represented by the sum of the useful signal S (x) and the noise n(x), or

$$f(x) = S(x) + n(x) \qquad (7)$$

The useful signal can be understood as the sum of the normal field, regional and local anomalies (residual - in case of using formal mathematical methods - Odstrčil 1989: 205-213). The

noise occures due to methodological, measuring errors, etc. In solving tasks from the field of archaeogeophysical prospection the studied territory has, as a rule, smaller dimensions (about 1 - 5 ha), therefore the normal field (Tn) can be considered virtually constant, and in the primary procedure it can be calculated from the measured data, e.g. in a statistical way - the maximum of the Gaussian curve of the distribution of frequencies (normal relative field). By regional anomalies are understood those evoked/generated by geological sources in great depths and/or inhomogeneities in smaller depths whose dimensions are considerably larger (5 and more times) than the size of the sought objects. Local anomalies are due to near-surface inhomogeneities with dimensions comparable to the searched for archaeological structures

$$f(x) = S(x)_{norm} + S(x)_{reg\,a} + S(x)_{loca} + n(x) \qquad (8a)$$

by subtracting the normal relative field formula (8a) will be in the form

$$f(x) = S(x)_{rega} + S(x)_{loca} + n(x) \qquad (8b)$$

Spectral characteristics, regional, local and of noise are given in Fig. 12.

Characteristic of regional anomalies are the largest wave lengths (the smallest frequencies), local anomalies have small wave lengths (medium frequencies), the noise spectrum includes all frequencies. In the program for separating residual anomalies from the measured data against the background of regional anomalies differences in their spectral characteristics are utilized.

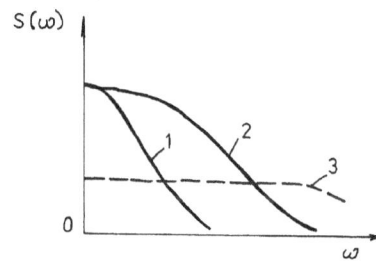

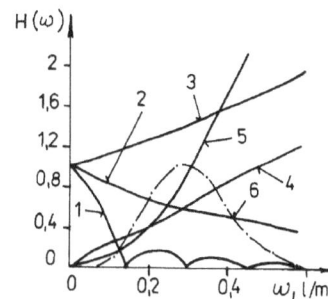

*Fig. 12 (Top).* *Relative spectral characteristics of a regional (1), a local (2) anomaly and noise (3) according to Nikitin 1986*

*Fig. 13 (Bottom). Frequency characteristics of fundamental transformations 1 - smoothed, 2 - analytical continuation at a higher level, 3 - analytical continuation to a lower level, 4,5 - calculation of higher derivatives, 6 - calculation of derivatives on a higher level (Nikitin 1986)*

The transformation formulas are based on the theory of potential fields. They make it possible to calculate from the values of the component of the field set at a certain level other components, potential, gradients, higher derivatives, to transform the fields on a higher or a lower level, etc. (Hašek - Segeth - Vencálek 1990: 156-192). The analysis of amplitude-frequency characteristics (Fig. 13) of these transformations makes it possible to choose those that increase the local anomalies and suppress the regional ones.

The conversion of the field to a higher level and the calculation of the mean value in the window highlights the low frequency components and reduces the high frequency ones. The conversion of the field to a lower level and the calculation of higher derivatives, on the other hand, highlights the higher frequencies, but it does not lower the low frequency ones, i.e. anomalies extended along the profile. The calculation of higher derivatives, after the conversion of the field to a higher level, allows the separation of frequency components in the range delimited from the sides of low and high frequencies. However, since they are measured at the nodes of the regular network, or they are interpolated into the network, the field is set discretely, and also its transformations are made within the amplitude-frequency characteristics of transformations from the above ones. These characteristics agree to a certain frequency, but then they either converge to zero or they oscillate about some constant. Thus the results of transformation of the field are distorted. In relation to the measurement errors such characteristic is, on the other hand, advantageous, because the high frequency components of the fields prevailing in the spectrum of the measuring errors are depressed.

For purposes of archaeogeophysical prospection a system of programs has been compiled allowing the separation of the high frequency component of the spectrum, and/or its individual components of the respective frequency of the field or operations not affecting this spectrum. In the former case we speak about filtering, in the latter about the transformation in the basic meaning of the word (Pašteka 1977:80-88).

They were above all
- the centering of the data measured
- the calculation of higher derivatives (for increasing the resolutions of the anomalies of the magnetic field),
- the calculation of derived anomalies of the field of weighted running averages (Griffin 1949: 39-56), etc.

The objective of the performed separation is to obtain geophysical effects of the object from the measured and subsequently processed data. The other components of the field are considered a sign of a deeper geological structure, stratigraphic or facial changes in the rock massif, and/or accidental measuring errors (Odstrčil 1989: 205-213).

The derivation of the filtering coefficients permitting the smoothing of the measured data and the calculation of higher derivatives (A.A. Lyubimov - G.A. Lyubimov 1983: 130-136) is based on the following consideration: If the values of the field f(x) set on a profile with a constant distance $\Delta x$ in

the running window having K nodes are approximated by the polynomial $\varphi(K) = ax^3 + bx^2 + cx + d$ by the method of least squares $F = \Sigma [f(x) - (x)]^2 = $ min., then in the centre of the windows the smoothed value of the field is numerically equal to the absolute member of the approximating polynomial, the first derivative to c, the second one to the double of parameter b and the third one to six times parameter a. From the results of the solution of the system of normal equations the unknown magnitudes a, b, c, d are determined (Hašek - Segeth - Vencálek 1989: 156-192).

For practical calculation of the smoothed curves and the second horizontal derivative $\partial^2 f/\partial x^2 = f_{xx}$ on the profile the following formulae of 5 and 7 points are used for K = 5

$$f(0) = 1/10^4(-857 f_2 + 3429 f_1 + 4857 f_0 + 3429 f_1 + 857 f_2)$$

$$f_{xx} = 1/10^4 \Delta x^2 (2857 f_2 - 1429 f_1 - 2857 f_0 + 2857 f_1 + 2857 f_2) \quad (9\,ab)$$

$$K = 7$$
$$f(0) = 1/10^4(-925 f_3 + 1429 f_2 + 3333 f_0 + 2857 f_1 + 1429 f_2 - 952 f_3)$$

$$f_{xx}(0) = 1/10^4 \Delta x^2 (1109 f_3 - 714 f_1 - 952 f_0 - 714 f_1 + 1190 f_3) \quad (10\,ab)$$

where f(0).......................smoothed value in the centre of the window
$f_{xx}(0)$.............................second horizontal derivative in the centre of the window
$f_{-3}, f_{-2}, f_{-1}, f_0, f_1, f_2, f_3$ .....measured values of the field ($\Delta T, \rho_a$ etc. and the stepping distance $\Delta x$.

In the above way it is possible to find the coefficients for smoothing and higher derivatives also in the plane, according to the differently selected running window (Kxq). For the calculation of the second horizontal derivative, e.g. in the window (5x5) the following formula is used

$$f_{xx}(0) = 1/10^4 \Delta x \,(-245\,(f_{2,2} + f_{2,-2}) + 980\,(f_{2,1} + f_{2,1}) - 490\,(f_{1,-1} + f_{1,1}) + + 122\,f_{1,2} + f_{1,-2}) + 245\,(f_{0,2}) - 980\,(f_{0,1} + f_{0,-1}) - 1388\,f_{0,0} - 694\,(f_{1,0} + f_{1,0}) + + 1388\,(f_{2,0} + f_{2,0}) + 980\,(f_{2,1} + f_{2,-1}) - 490\,(f_{1,1} + f_{1,-1}) + 122\,(f_{1,2} + f_{1,-2}) - 245\,(+ f_{2,-2})) \quad (11)$$

where $f_{xx}(0)$.....second horizontal derivative in the centre of the window
$f_{-2, -2} ... f_{2, 2}$ ......measured values of the field ($\Delta T, \rho_a$ etc.) in the running square window.

Residual anomalies can be calculated as the difference of the measured field and the regional anomaly. If the regional field is expressed as a weighted mean value inside the planar domain S, the weight of the individual data depending on their position with respect to the centre of the domain in which the calculation point P is situated and if this magnitude is subtracted from the measured one, the residual anomaly of the field is obtained.

In calculating the regional anomaly it is started from the integral transformation defined by the formula (Mareš et al 1979: 58-59; Hašek - Vencálek 1989: 179-192).

$$\Delta T_{rga}(P) = 1/\mu_s \iint \Delta T(M) \mu M \, ds \quad (12)$$

where M............................an arbitrary point in

domain S

$\mu(M)$.................................the weight function in point M for domain S

$\mu = \iint \mu(M)\,ds$...................cofactor standardizing expression (12) so that function $\Delta T_{reg.a.}(P)$ be expressed by the weighed mean value in domain S.

For the practical calculation the weight function $\mu(M) = 1$, i.e. the values of the field are ascribed the same weight in all points and the domain S is in the form of a circle with radius $r = \Delta x(5)^{1/2}$ ($\Delta x$ is the step of measurement). The regional field can then be expressed as a function of the mean values of several circles of radius r. Thus, for the case of three circles chosen in such a way as to pass through the corners of the square network ($r = \Delta x, \Delta x(2)^{1/2}, \Delta x(5)^{1/2}$ :

$$\Delta T_{rega}(P) = 1/10\,(f_{0,0} + 1/2(f_{1,0} + f_{0,1} + f_{-1,0} + f_{0,-1}) + (f_{1,1} + f_{-1,1} + f_{1,-1} + f_{-1,-1}) + 3/8(f_{2,1} + f_{1,2} + f_{-1,2} + f_{-2,1} + f_{-2,-1} + f_{-1,-2} + f_{1,-2} + f_{2,-1}) \quad (13)$$

and residual anomaly

$$\Delta T_{reza} = \Delta T_{mer} - \Delta T_{rega} \quad (14)$$

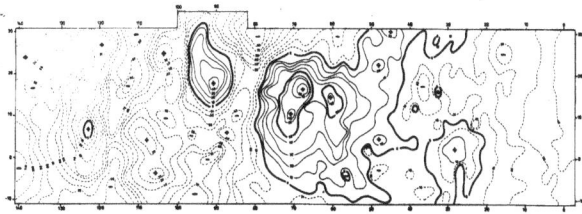

**Fig. 14.** *A map of isanomales $\Delta T$ from the space of the extinct medieval village of Srnávka near Svinošice, district Blansko*

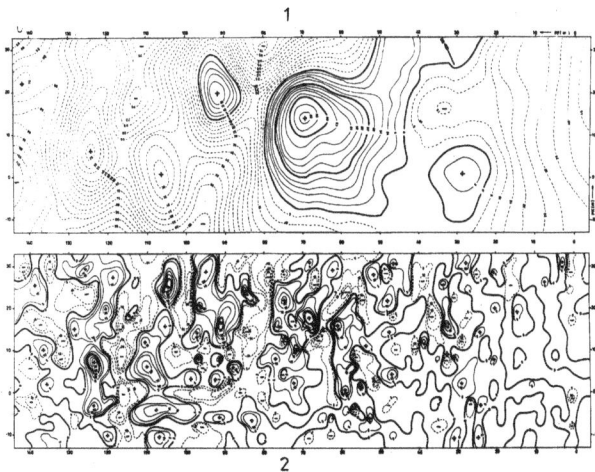

**Fig. 15.** *A map of centred (1) and residual(2) anomalies $\Delta T$ for $r = \Delta x(5)^{1/2}$ from Fig. 14*

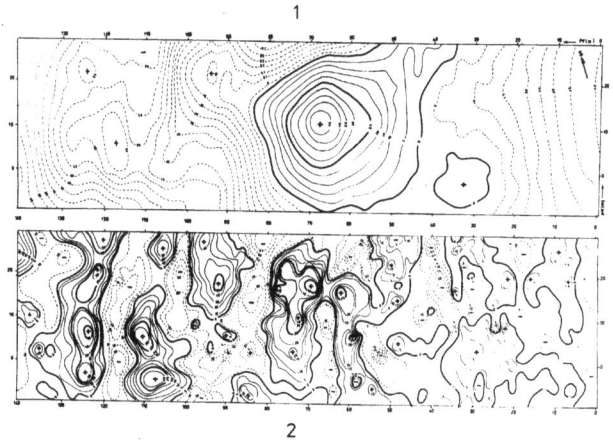

**Fig. 16.** *A map of centred (1) and residual (2) anomalies $\Delta T$ for $r = 2\Delta x(5)^{1/2}$ from Fig. 14*

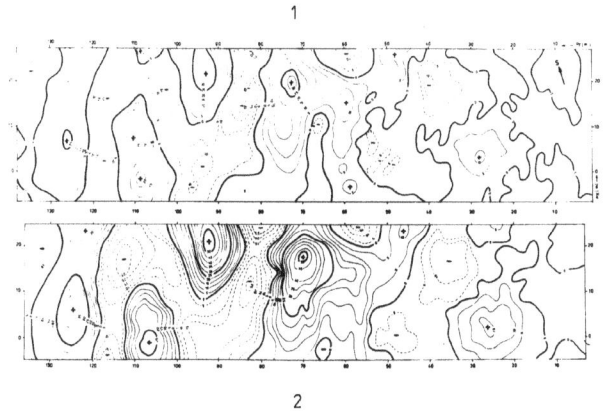

**Fig. 17.** *A map of residual anomalies $\Delta T$ according to Saxov-Nygaard (1953, 255-269) for $r = (1-3)\Delta x(5)^{1/2}$ (1) and $r = (1-2)\Delta x(5)^{1/2}$ (2) from Fig. 14*

In Figs. 14 to 17 there are comparisons of different radii r in calculating the centred and residual anomalies $\Delta T$ according to (13) and (14) from the region of the extinct medieval village of Srnávka near Svinošice (district Blansko). From the submitted pictures it follows that the effect of the employed filter in the processing depends on the ratio of the area of the planar domain and the anomaly due to the studied archaeological object.

From Fig. 15 it can be seen that if $r = \Delta x(5)^{1/2}$ is applied for the above case, in the map of residual anomalies $\Delta T$ prevailingly noise is separated. All information about the environment (the useful component) is contained in the map of "residual" anomalies. In larger r, such as $r = 2.\Delta x(5)^{1/2}$, $(1-3)\Delta x(5)^{1/2}$ and $(1-2)\Delta x(5)^{1/2}$ (Figs. 16 and 17) in the compiled maps, despite different accompanying anomalies of the opposite sign, are more conspicuously represented the local effects of the investigation against the background of "regional" anomalies with the sources of geological origin, probably in greater depths. It is, however, not possible to exclude a sign of extensive archaeological structures, such as a complex of shapes, layers of daub in combination with Fe objects, etc.

From the statistical approach to the measured data there starts the method of coefficients of mutual correlation which can be used in that case if the local anomaly is apparently funnelled in one direction. The method is based on the gradual comparison of effects of the inhomogeneity set in a certain window with the measured field (Hašek et al. 1986). As standard can be set the theoretical curve over the object searched for or the values of the field measured at places where the position of the studied structure was confirmed by another method. This process can also be applied to the derived fields. The output represents a matrix of coefficients of mutual correlation whose columns match the individual profiles. Max. values of the correlation coefficients whose location on the profile is always related to the centre of the sections searched for then determine the continuation of the studied anomaly also on the neighbouring profiles. The applied formula for the calculation of coefficient of mutual correlation (R) between the standard field e(x) set in the window of K nodes and a section of the profile i with the central point j.

$$R_{ij} = \frac{\sqrt{(1/K \sum_{j=-(K-1)/2}^{(K-1)/2} e_j \cdot x_{ij}) - \bar{e} \cdot \overline{x_{ij}}}}{\sqrt{(1/K \sum_{j=-(K-1)/2}^{(K-1)/2} -\bar{e}^2)^{1/2} - \bar{e}^2} \sqrt{(1/K \sum_{j=-(K-1)/2}^{(K-1)/2} x_{ij}^2)^{1/2} - \bar{x}}}$$  (15)

where $R_{i,j}$... coefficient of mutual correlation between the standard field and the section of the profile with central point j in the window of K nodes.

$e_j$......standard field set at points $-(K-1)/2 \ \Delta x$, $-(K-1)/2 + 1 \ x ...0 ... (K-1)/2-1 \ \Delta x$, $(K-1)/2$, $\Delta x$

$\bar{e}$......mean value

$x_{i,j}$.... values of the measured or derived field on the profile i at points $x_i$, $(-K-1)/2\Delta x ...x_{i,j}...x_i$, $(K-1)/2\Delta x$

$\overline{x_{i,j}}$ ...mean value of the field on profile i in the window of K nodes with the centre at point j.

A practical example of the correlation of the Neolithic circular structure near Němčičky (district Znojmo) by means of this method (K=5) is given in Fig. 18 (cf. Hašek - Měřínský 1991: 106-107).

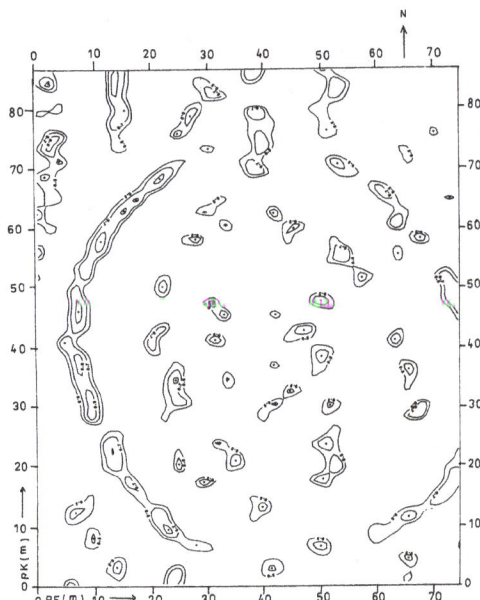

**Fig. 18.** *A map of isolines of coefficients of mutual correlation from the locality of Němčičky, district Znojmo*

## I. Resistance Profiling

For an appropriate choice of the geometry of electrodes for the location of the studied objects it is necessary to follow the depth range depending not only on the arrangement of the electrodes (distance AB = 2l), but also the dimensions of the archaeological body and the resistance conditions. In a homogeneous environment it can be understood as such depth h in which the current density equals at least one-tenth of the current density $j_0$ on the Earth's surface.

The current density $j_h$ below the centre of the electrode arrangement is given by the well-known relation (Dachnov 1951)

$$j_h = j_0(1/(1+h/l)^2)^{3/2}$$  (16)

where l is the half-distance of the current electrodes.

The dimension of arrangement 1 is understood as the depth range. From expression (16) it is evident that the current density is dependent on the ratio h/1. Thus, at h = 6 l (2 l = 1 m, h = 3 m) $j_h$= 0 no information can be added about the studied object at the above depth by the employed electrode geometry and, on the contrary, at 2 l = 3, the resistance data about the body studied can be found according to (16) to the depth h = 1.5 m. For practical evaluation of changes in resistance conditions with the depth it is recommended to carry out RP with several depth interventions and/or the measurement of VES.

a) Nonconducting Objects

Archaeological structures (recessed objects, destruction positions, etc.) have mostly the shape of horizontal platforms. This position is, however, not very suitable for the indication of nonconductors, because no major distortion of the current lines flowing around the object occurs nor is there a great inflation of equipotential lines. If, however, those bodies are thicker, there is an inflation of the density of equipotential lines above them. The expression is the measured high apparent specific resistance. The indications are particularly noticeable in an environment with low resistance.

Combined profiling, and/or three-electrode (potential, gradient) arrangement registers a thick nonconductor in all cases by a considerable anomaly $\rho_a$, also near the conductor. A thin nonconductor shows well too. On the curves of the Wenner profiling with the perpendicular arrangement of electrodes, the nonconductor, thicker in comparison with the distances of electrodes, manifests itself by a simple maximum. A thin conductor, on the contrary, cannot be located by this arrangement. The dipole equatorial arrangement is unsuitable for searching for nonconductors. By means of the dipole axial arrangement two equal maximums are measured on each curve, corresponding to the passage of the two dipoles over the nonconductor (see Fig. 35). If the resistances are related to the two dipoles, two curves are obtained whose maximums will be common over the nonconductor.

The arrangements of electrodes at which the current lines are parallel to the searched for nonconductive body are not suitable for their mapping.

Finding sufficiently large resistance inhomogeneities due to man's activity by means of RP (and/or VLW-R, DEMP) does not cause any trouble. The expression of a thin nonconductor (relics of foundation masonry) at $\rho_a$ curves is, however, less evident than that of a thin conductor and can be concealed by secondary indications. Thus, dipole axial profiling has very complicated curves over arbitrary inhomogeneities and it is unsuitable for the location of thin nonconductive positions. Similarly, also at KRP and SRP curves there are indications due to the passage of electrodes over a resistance inhomogeneity. The most suitable for finding thin nonconductors seems to be, for the time being, the method of the central gradient with line electrodes.

b) Conductive Objects

For the location of thick and thin plate shaped conductors which, however, are a less frequent case in archaeogeophysical prospection, as optimum appear combined profiling, dipole axial and/or equatorial profiling. The searched-for body (such as a moat, a recessed object, a grave pit, etc.) displays itself by crossing the two branches of the resistance curves. If the direction of the conductive formation is known, then the Wenner profiling, the method of the central gradient with the perpendicular arrangement of electrodes and a combined central gradient. These arrangements grant, at the correct orientation of profiles, simple and easily interpretable curves (Fig. 19) even under complex geological conditions.

| GROUP | PROFILING METHOD | TYPICAL RESISTANCE CURVES OVER A CONDUCTIVE BODY |
|---|---|---|
| symetrical profiling | Wenner's AMNB | |
| | Schlumberger's A MN B | |
| | center gradient A MN B | |
| non-symetrical profiling | combinated A MN , MN B | |
| | dipole axis AB MN , MN AB | |
| | combinated center gradient A MN , MN B | |
| perpendicular setup profiling | Wenner's perpendicular setup (A MN B) | |
| | center gradient with perpendicular setup (A MN B) | |
| | dipole equatorial (AB)(MN),(MN)(AB) | |

*Fig. 19. A list of main modifications of RP and a typical resistance curve above a plate-shaped conductive body according to Karous (Mareš et al. 1983)*

There exists a dependence of the RP results on the direction in which the measuring system crosses the archaeological object. If the environment is homogeneous, then the measured $\rho_a$ does not depend on the orientation of the arrangement of electrodes. With the occurrence of an archaeological object changing the distribution of the current, $\rho_a$ depends on the position of electrodes with respect to the object, on the ratio of the resistance of this structure and on the ambient environment. In places with archaeological objects the measured values of $\rho_a$ will change in dependence on the orientation of the geometry of electrodes (and/or profiles) for the extension of the body searched for. For the above reasons also different magnitudes of $\rho_a$ will be measured in both selected directions.

c) The Effect of Relief

Also in archaeogeophysical prospection the results of RP can be affected by the relief of the field without the possibility of quantitative corrections (Gruntorád-Karous 1972). An inclined slope with a planar surface has practically no effect on curves $\rho_a$ They are affected only by a sudden change in the field slope. In such a case they can be affected by error even greater than 10 %. The $\rho_a$ curves are the least affected if the arrangement of electrodes is parallel to the elongated morphological elevation - defence line-vallum and/or depression- moat, etc. On their peak the found magnitude $\rho_a$ will be greater thanks to the increased density of the current lines, unlike the valley, where, on the contrary, a lower value can be measured. A more complicated effect is that of structures oriented perpendicular to the profiles and the measuring systems. In general it can be said that the elevated structure will appear as a conductor, the depressed one as a nonconductor. The effect of relief is evaluated only quantitatively in that case, i.e. in the interpretation anomalies due to these formations are not taken into consideration.

d) Data Procedure

Some of the programs discussed in Chapter 3 can also be used for processing data from the method of resistance profiling (similar also to the VLW-R and DEMP methods). It is above all the centreing of the profile curves of apparent specific resistances and/or also the calculation of higher derivatives. On the statistical approach to the solution of the task are based e.g. the method of amplitude filtering, controlled filtering and the above method of coefficients of mutual correlation (see Chapter 3.3.1.1.).

The separation of "anomalous zones" from the overall course of $\rho_a$ curves originates in J. Grodnicki's work (1977: 58-79). The principle is the processing of the measured data fulfilling the criterion of the normal distribution by the so-called method of amplitude filtering. For the separation of anomalies of $\rho_a$ is first of all made the calculation of the selection average of $\rho_a$, of the variance of $\sigma^2$ and the standard deviation $\sigma$. The following formulas calculate extreme values at the measured points x:

$$E_x = -\sigma^2 + (\rho_a - \bar{\rho}_a)^2 \qquad (17)$$

or

$$\rho_a - \sigma > \rho_a^{anom} > \bar{\rho}_a + \sigma \qquad (18)$$

The positive extreme values of $E_x$ locate the position and approximate width of anomalous bodies with respect to the surrounding environment (Hašek et al. 1983, 1984a). A practical example of the application of the above method in the space of the Roman camp Leanyvár near Iža (district Komárno) at the location of the fundament masonry of stone objects is given in Fig. 20.

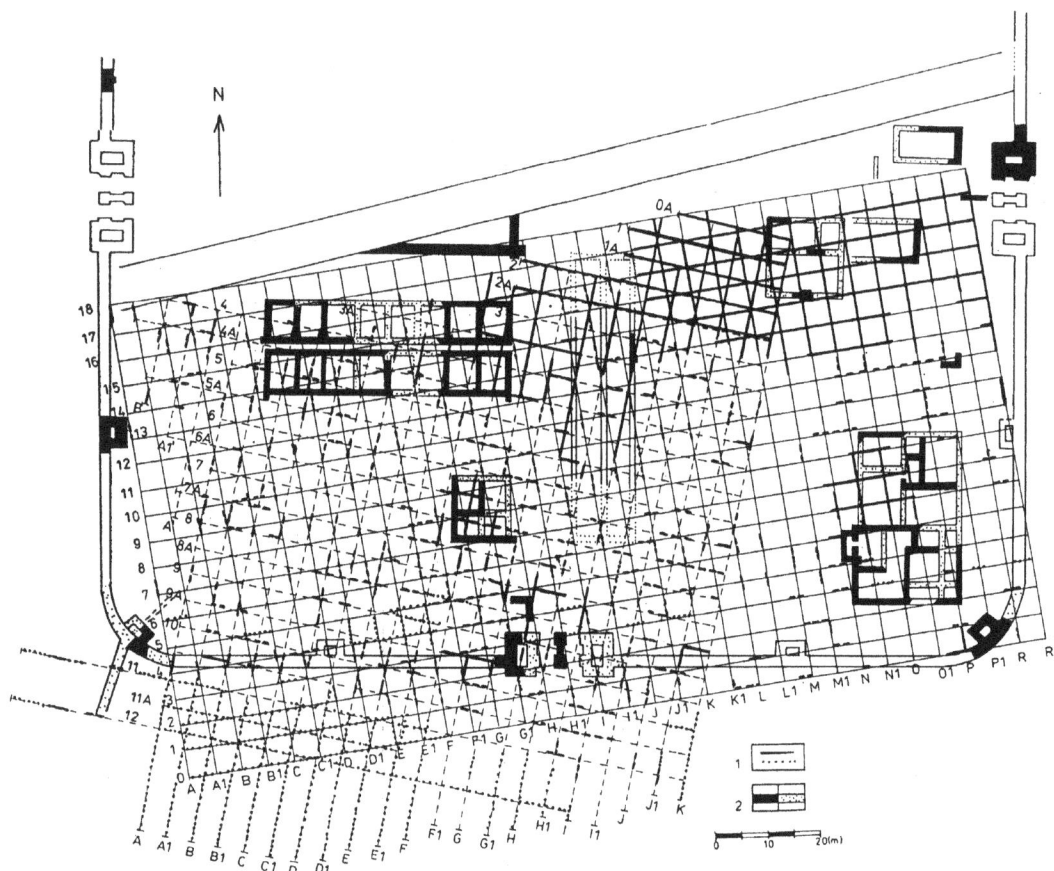

*Fig. 20.* An example of the separation of "anomalous zones" $\rho_a$ in the space of the Roman camp Leanyvár near Iža, district Komárno 1 - positions of interpreted nonconductive and conductive zones from $\rho_a$ curves, 2 - originally exposed foundation masonry of structures or their negative imprint

The algorithm of controlled filtering for a complex of two geophysical methods, such as OP with two depth interventions, DEMP - microgravimetry etc. is based on the calculation and comparison of correlation coefficients for anomalies and interferences within the selected window of N profiles and m points of measurement at each profile (Sikorskiy 1979: 120-126).

Every accidental function n, in accordance with (7), consists of the useful signal S (z,x) and the interference n (y,x). That is,

$$f_{1,2}(y_k, x_i) = S(y_k, x_i) + n(y_k, x_i) \qquad (19)$$

where 1,2 .......indices of geophysical methods
k ...................profile number
i ...................number of meters

The separation of the useful signal and the interference is done by the summation along the profiles within the window for each method

$$S_{1,2}(y, x_i) = 1/N \sum_{K=1}^{N} f_{1,2}(y_k, x_i) \qquad (20)$$

$$n_{1,2}(y, x_i) = f_{1,2}(y, x_i) - S_{1,2}(y, x_i) \qquad (21)$$

(The calculation is related to the central profile of the elected system N = 3).

The linear dependence of the useful signal of two applied methods can be estimated by means of coefficients of correlation

$$R_s = 1/((m-1)\sigma_{s1}\sigma_{s2}) \sum_{i=1}^{m} (S_1(y, x_i) - \bar{S}_1)(S_2(y, x_i) - \bar{S}_2) \qquad (22)$$

$$\text{where } \bar{S}_{12} = 1/m \sum_{i=1}^{m} S_{12}(y, x_i)$$

$$\delta S_{12} = (1/(m-1) \sum_{i=1}^{m} (S_{12}(y, x_i) - \bar{S}_{12})^2)^{1/2}$$

Analogously for interferences

$$R_n = 1/[(m-1)\sigma_{n1}\sigma_{n2}] \sum_{i=1}^{m} (n_1(y, x_i) - \bar{n}_1)(n_2(y, x_i) - \bar{n}_2) \qquad (23)$$

where

$$n_{1,2} = 1/m \sum_{i=1}^{m} \bar{n}_{1,2}(x, y)$$

$$\sigma n_{1,2} = (1/(m-1) \sum_{i=1}^{m} (n_{1,2}(y, x_i) - \bar{n}_{1,2}))^{1/2}$$

The reliability of the separation of geophysical anomalies is based on the estimate of coefficients of the correlation of the useful signal. The dimension of the window is usually chosen so that one starts from the mean dimensions of anomalies. If a great correlation dependence of useful signals exists and a small one in interferences, it is possible to speak about the correlation of anomalies.

The solution of the case of existence of anomalies originates from the analysis of the parameter

$$\mu = R_s - R_a \qquad (24)$$

In Fig. 21 there is an example of the comprehensive evaluation of the DEMP method and microgravimetry in solving the issues of cavities in connection with old mining activity at Zbýšov near Brno by using the method of coefficients of mutual correlation, controlled and amplitude filtering. The position of the gallery is very clear from the results of the processing (Hašek - Měřínský 1989: 180-181).

## II. Vertical Electric Sounding

The accuracy of the quantitative processing of VES curves depends to a considerable extent on the qualitative interpretation on the basis of which it is possible to get an overall idea of the changes in the geoelectric section and its parameters in the studied archaeological structure and determine the linkage of this section with the geological structure of the topmost layers of the territory of interest.

Isoohmic sections $\rho_a$ (x,h) (h = r/2 = AB/4) and the maps of isoohms $\rho_a$ (x,y) for the given distances r are utilized in archaeogeophysical prospection mainly for monitoring the deposition of the layers (see Fig. 49) and the clarification of the main regularities and/or changes in the studied geoelectric section (thickness, depth of the base of the cultural layer, lithological indentation of the moat fillings, the homogeneity of the massif with the occurrence of cavities, etc.). The expression of the individual geoelectric layers depends in general on their thicknesses and resistance properties. That means that the greater the thickness of the studied horizon, such as a cultural layer, the nearer it is to the surface, the more different their resistances from the overlying (gravel, loess) and underlying (clay, sand and other) rocks, and the better it will appear in the results of quantitative interpretation. With the growing depth and specific resistances the rate of accuracy as well as the ability of its separation drops, if no complementing data from excavation, probes, engineering-geological bores etc. are available.

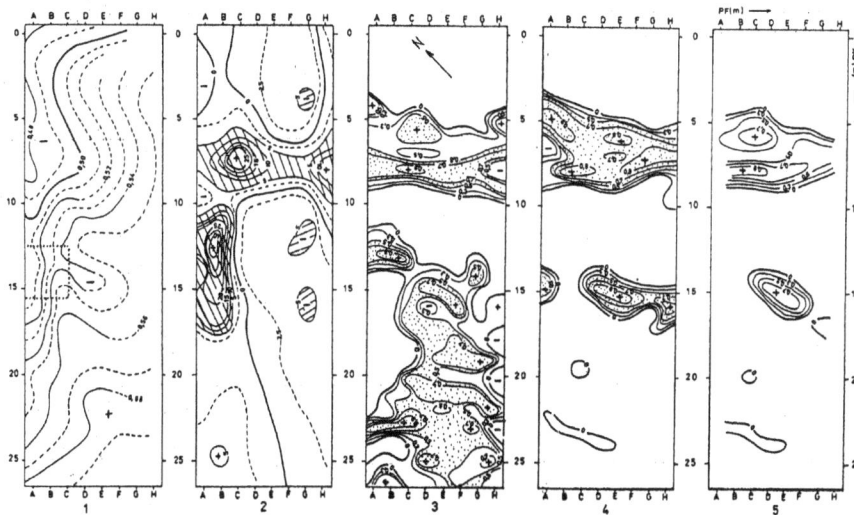

*Fig. 21.* *Comprehensive processing of geophysical data at the locality of Zbýšov near Brno, 1 - map of relative anomalies $\Delta g$ ($\mu ms^{-2}$), 2 - map of isolines of anomalous resistances from the DEMP method (ohmm), 3 - map of isolines of the coefficient of mutual correlation for $\rho_{DEMP}$ (K=3), 4 - map of isolines of coefficients of mutual correlation for $\Delta g_{rel}$ (K=3), 5 - map of isolines of parameter $\mu$.*

## III. Dipole Electromagnetic Profiling

The possibilities of determining the resistance properties of objects of anthropogenic origin by means of the DEMP method with instruments EM-31, KD-1 and EM-38 follow from the dependence of the measured components of the electromagnetic field on the apparent specific resistance of the equivalent semispace. These properties become considerably simpler on fulfilling the condition of a near zone.

$$N = R/\delta = (\omega \sigma \mu_0/2)^{1/2} \quad r \ll 1 \qquad (25)$$

when the depth of the penetration (skin) $\delta$ considerably exceeds the distance r between the transmitter (Tx) and the receiver (Rx) (m), as well as the thicknesses of the horizontally layered semispace. On the above assumption the curves of detectability of extensive horizontally deposited conductive and nonconductive bodies are easily interpreted, which can be utilized in the interpretation of results in archaeogeophysical prospection. The situation is more complicated in measuring minor anomalous bodies whose horizontal dimensions are smaller than the distance r between Tx and Rx. In that case nonconductive inhomogeneities can be located with great difficulties. The interpretations are considerably complicated and ambiguous even in conductive objects (Záhora 1989: 226-252).

The measurement of apparent conductivity ($\sigma_a$) at three height points above the ground allows the determination of three searched parameters of a simple two-layer environment. A certain redundance of information corresponding to the combination of the found data at ZZ and YY polarization (6 measured values) is only necessary for the corresponding processing of the measured data (a credible assignment of these values to theoretical curves). The above conductometers are mostly used for quick profile measurements.

In processing and interpreting the results of the DEMP method the apparent conductivity ($\sigma_a$) determined at the individual points (2) is first transferred to apparent resistances ($\rho_{DEMP}$) (3) which then, according to suitable chosen weight coefficients for the neighbouring points, can be centred and anomalous resistances are determined ($\rho^{DEMP}_{anom}$). From the smoothed profile lines maps of isolines $\rho_{DEMP}$ or $\sigma_a$ and $\varsigma^{DEMP}_{anom}$ are plotted (see Chapter 3.3.1.). For further processing of profile curves, particularly in searching for elongated line inhomogeneities (2 1/2 D) it is possible to utilize the method of coefficients of mutual correlation and/or in the application of two methods by controlled filtering (see Fig. 21).

In Fig. 22 there is the processing of measurement from the DEMP method at the locality Předklášteří - Tišnov II from which it was possible to assume the existence of stone foundation masonry of a rectangular object. On investigation a minor sacral building was found - a chapel (Belcredi 1993: 323-340).

For the evaluation of profile measurements $\rho_{DEMP}$ ($\sigma_a$) or $\partial T/\partial Z$ it is also possible to use the statistical method of main components (Hašek et al. 1986), even though its chief assertion is in the qualitative interpretation of results at a number of methods applied by which mostly different or equal physical magnitudes are found.

## IV. The Resistance Version of the Method of Very Long Waves

The depth range of the VLW method depends on the depth of penetration of the electromagnetic wave which can with sufficient accuracy be expressed by the relation

$$p = 500(\rho/f)^{1/2} \qquad (26)$$

where $\rho$ .... specific resistance of the environment (ohmm)
f ... frequency (Hz)

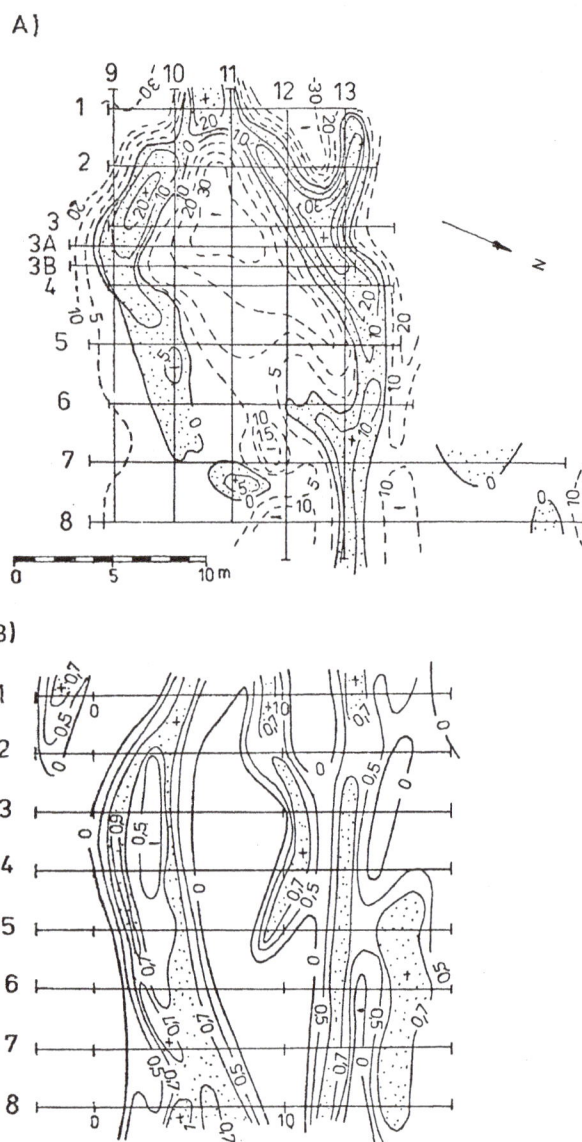

*Fig. 22.* A map of $\rho_{DEMP}$ isolines and of coefficient of mutual correlation from the locality of Předklášteří - Tišnov II,.

From formula (26) it follows that the depth range is directly proportional to the root of the resistance of the environment and inversely proportional to the frequency of the primary field. As stated in Chapter 3.1.2.2., most transmitters operate in the VLW band at the frequency of 15.1 to 22.16 KHz. The depth differences due to the change in frequency of the transmitter are not apparent. More substantial differences occur due to the change in resistance of the environment. If the penetration of the wave is about 30 m at usual frequencies of the transmitting stations and the resistance of the environment 100 ohmm, then at resistances of about 10 ohmm the depth range will be about 10 m.

The interpretation of resistance measurements of the VLW method is similar to the processing of usual resistance measurements by means of DC. However, there exist some differences, because the physical substance of the measurement is different. Here belong
   a) a different character of primary field

b) the dependence of the depth range on the frequency of the transmitter and the resistance of the environment,

c) the dependence of the measured values on the mutual orientation of the transmitter and the direction of the structures,

d) the possibility of utilizing the phase shift between the electric field and the magnetic one ($\varphi$) for the assessment of the geoelectric section.

In the process of choosing the arrangement it is necessary to consider the shape of the curves above simple bodies. The E arrangement is more advantageous for searching for conductive bodies, H for searching for nonconductive positions. In the location of contacts these issues are somewhat more complicated. More apparent indications arise at the H-polarization. The course of the curves is, however, more complicated. In mapping archaeological structures it is therefore more advantageous to start from the information about the overall geological structure of the territory. In a simple space it is more advantageous to use the H arrangement, in a more complicated one the E arrangement.

From physical assumptions it follows that above a homogeneous semispace the phase shift $\varphi = 45°$ and the measured specific resistance $\rho^{VLW}$ is the specific resistance of the environment. If the measurement is made over a two-layer environment, where the thickness of the first layer is smaller than the penetration of the electromagnetic wave, the value of $\rho^{VLW}$ will be affected by the resistance of the second layer. Then $\rho_2^{VLW}$ will be measured. The magnitude of the phase shift will differ from 45°. In the case $\rho_1 < \rho_2$, the value of $\varphi = 0-45°$, in the opposite case, when $\rho_2 < \rho_1 = 45-90°$. The phase shift thus yields qualitative information about the geological section. Under favourable conditions the parameters $\rho^{VLW}$ and $\varphi$ can also be used to determine the thickness of the first layer.

In the interpretation of anomalies $\rho^{VLW}$ above the vertical contact it is necessary to distinguish whether it is H or E polarization. For theoretical curves above the contact starting towards the surface it holds that at E-polarization the change in impedance and thus also in resistance is continuous. At H-polarization the apparent specific resistance changes with a jump. If there are cover formations of great thicknesses above the contact, $\rho^{VLW}$ changes continuously also in H-polarization.

Amplitudes of anomalies of resistance and phase curves in plate-like bodies depend on the differentiation of the conductivity of the conductor and of the ambient environment (Karous 1982: 77-79). At E-polarization the amplitude of resistance curves increases with increasing conductivity of the local conductive inhomogeneity. In phase curves the broad minimum of the curves is deepened, whereas the relative size of the local maximum does not change very much. Thicker conductive positions appear particularly if they have a relatively greater conductivity in the phase curves $\varphi_E$ by two extremes that can be interpreted as two parallel conductive plates. To a wider conductive zone corresponds a wider minimum of the curve $\rho_E$ with one local extreme. Phase curves have large minimums above the horizontally located conductive positions when, although the amplitude of the anomaly of the phase shift is small, the deeper deposited conductor is expressed by an extensive anomaly. In the case of horizontal deep located conductive bodies, the two arrangements are approximately equivalent.

The processing of the results of measurement of the VLW-R method can be performed according to the methods discussed in Chapter 3.1.2.1. "Resistance Profiling" and/or DEMP. The methods are:

a) centering of the measured data and the calculation of higher derivatives,

b) making maps of isolines of $\rho^{VLW}$,

c) calculation of coefficients of mutual correlation,

d) the method of amplitude filtering,

e) the method of controlled filtering with two applied methods.

*V. The Geophysical Radiolocation Method*

The evaluation of the depth ranges with the potential possibilities of the radiolocation method for the purposes of locating near-surface anthropogenic objects was dealt with by R. Záhora (Hašek et al. 1983a). The author gives frequency characteristics of the specific resistance and the relative $\varepsilon_r$ in a relatively wide band of the amplitude spectrum of transmitted pulses of the apparatus SIR-7 found by laboratory measurement of rock samples at different degrees of water saturation within the limits 0 to 1.

These curves exhibit a substantial reduction of the specific resistance with increasing frequency and with water saturation which causes the corresponding increase in the inhibition of electromagnetic energy in the given environment. The drop in the specific resistance at f = 80 MHz below the value of 20 ohmm causes the inhibition coefficient to increase to the value equal to or grater than 24 dB/m, to which corresponds the theoretical range of the method to 1.8 m. Calculations of the depth ranges for different loss environments agree well with the experimentally verified ranges obtained either in the course of experimental measurement, such as by means of the apparatus SIR-7 and/or with depth ranges given by the manufacturer. From these sources it generally follows that e.g. for the specific resistance of the environment of 125 ohmm (zero or very low frequency) the depth range to a metal plate is 2.1 m. It is possible to state that near-surface inhomogeneities in the Quaternary can be found by the radiolocation method (band middle about 80 MHz) to very small depths, about up to 2 m, max. to 3-4 m depending on the resistance conditions of the locality studied. For the detection of small metal objects it is necessary to use very short pulses (1-2 ns). The middle of their frequency band is relatively large (500 MHz). As most rocks exhibit considerable inhibitions for this frequency band, it is possible to detect the above bodies to the depths of order of 0.5 - 1.5 m, which is sufficient for purposes of archaeogeophysical prospection. The dimensions of the objects searched for should not be smaller than the wave length, corresponding to the middle of the frequency band in the rock.

### 3.3.2. Quantitative Interpretation

The result of the quantitative interpretation of the measured data is a physico-archaeological model of the locality whose parameters are the physical properties of objects, their shape, size etc. It is also possible to find the type of the archaeological model, such as a fortification system, a habitation object, a metallurgic appliance, etc. (Hašek - Segeth -Vencálek 1990: 156-192; Hašek - Měřínský - Segeth 1990: 27-34, etc.).

The completeness of the model depends on many factors. Among objective factors there belong the differentiation of physical properties of objects of our interest with the rock environment, their size, the density of the measurement network, its accuracy, potency of noise, measurement by means of a complex of methods, etc.

At the present time quantitative interpretations are carried out for the purposes of archaeogeophysical prospection particularly in the magnetometric method and the VES curves.

### 3.3.2.1. Magnetometric and Geoelectric Modelling

In order to obtain a more detailed set of data both for the determination of the methodology of archaeogeophysical prospection, such as the choice of the measuring point density with respect to the size of anomalous bodies, the choice of the optimum geometry of electrodes in the RP method from the point of view of the size and width of the amplitude response, the depth range, etc., and also for the credible interpretation of the measured data (a sign of the artifact in the measured field according to the size, depth and physical property with respect to the ambient rock environment) it is advisable to carry out the so-called modelling which can be implemented by a numerical calculation or by physical measurement (Hašek - Měřínský1990: 50-68; Halíř - Hašek 1989: 193-204; Hašek - Měřínský1989: 45-53).

Mathematical modelling is the deduction of theoretical geophysical characters of simplified artifacts of an archaeological locality by means of numerical methods. Anthropogenic objects, such as a furnace, cavity, moat, defence line - vallum, recessed hut, tomb, etc., are replaced by simple geometrical two-dimensional and three-dimensional models (a circle, a cylinder, a prism, a polyhedron) with given dimensions, depths of deposition and physical properties (magnetic susceptibility, remanent magnetization, specific resistance, conductivity) located in the semispace which itself has its physical properties (Hašek - Segeth - Vencálek 1990: 156-192). The modelling is used above all in solving direct tasks of the individual geophysical methods, when theoretical anomalies are calculated above the body or the environment. The sum of models simplifies and simulates the actual sources of the individual geophysical anomalies, including the surrounding field. Further, in some methods the solution of the reversed task is applied whose role it is, on the basis of measured data to quantitatively and qualitatively describe unknown sources and the environment surrounding them. From time to time mathematical modelling is utilized for deducing the filter and operator parameters serving for the suppression of noise and uncalled-for anomalies in separating of the anomalies of the searched for archaeological objects from the integrated geophysical field.

The coherence between the measured geophysical field and the result of mathematical modelling need not yet be a proof of identity of the actual sources and partial models, because the interpretation of the measured data in most geophysical methods is not unambiguous (the task is incorrect) and the reality cannot always be described by simple models. For reducing this ambiguity complementary information from excavation, boring and other works is utilized.

Mathematical geophysical modelling is made on computers with an interactive approach (Hašek - Měřínský1990: 52-55), when the interpreter can operatively interfere with the choice of methods, models and their parameters and immediately compare the results of modelling with the measured data and obtain profiles and maps of isolines as a graphical output. The compiled programs allow in several interactions the deduction of parameters of the given or automatically chosen shape searched for, such as the minimalization one by the Marquardt method (Marquardt 1963: 431-441) and/or its combination with further methods.

Physical modelling is a complex of methods of contributing to and measuring anomalies of artificial physical fields evoked by bodies (models) made of corresponding materials located to the environment with constant or variable physical properties. Natural conditions are simulated by this model of the given scale. Physical modelling is considerably disposable and permits quick combinations of parameters of models and the variability of the methodology of measurements and apparatuses.

Whereas in magnetometric measurements mathematical modelling is implemented, in geoelectric methods mathematical and physical modelling is combined, such as in the electrolytic trough.

### Modelling the Magnetic Field

Mathematical modelling of the magnetic field of simple bodies is used to determine the course of magnetic anomalies of the archaeological objects searched for (direct task), as well as for estimating the parameters of sources of the measured anomalies (reversed task). So far a considerable number of mathematical and numerical relations have been published for the calculation of the direct task. First, magnetic effects of simple homogeneous bodies ($\Delta Z$, $\Delta H$, $\Delta T$) were deduced, now the authors solve magnetic effects of inhomogeneous bodies as well as objects of more complicated shapes, such as Golcman et al. (1981): Golcman-Kalinina (1983), etc.

In mathematical relations for the total intensity of the magnetic field $\Delta T$ used for the solution of the direct task (Lindner-Scheibe 1978: 29-45) archaeological objects are assumed to be magnetized by the geomagnetic field, the other sources of magnetization being zero towards it.

The geomagnetic field is described by the vector of total geomagnetic intensity $T_o$ with inclination I. The projection of vector $T_o$ into the x,y plane is the horizontal component $H_0$, situated in the direction of the local magnetic meridian (M.M.). The system of coordinates is oriented in such a way that the axis of the body is parallel to y-axis and with the local M.M. it makes the angle (Fig. 23). Magnetic susceptibility (æ) is set in SI units, C is the bearing, i.e. the angle between the magnetic north and the x-profile.

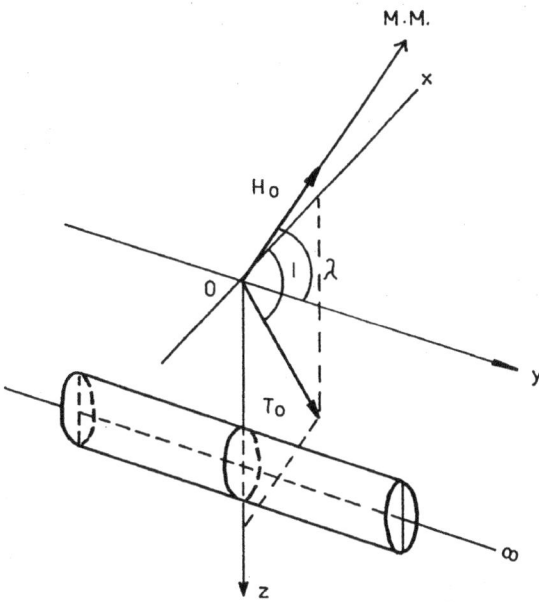

*Fig. 23.  Elements of the geomagnetic field*

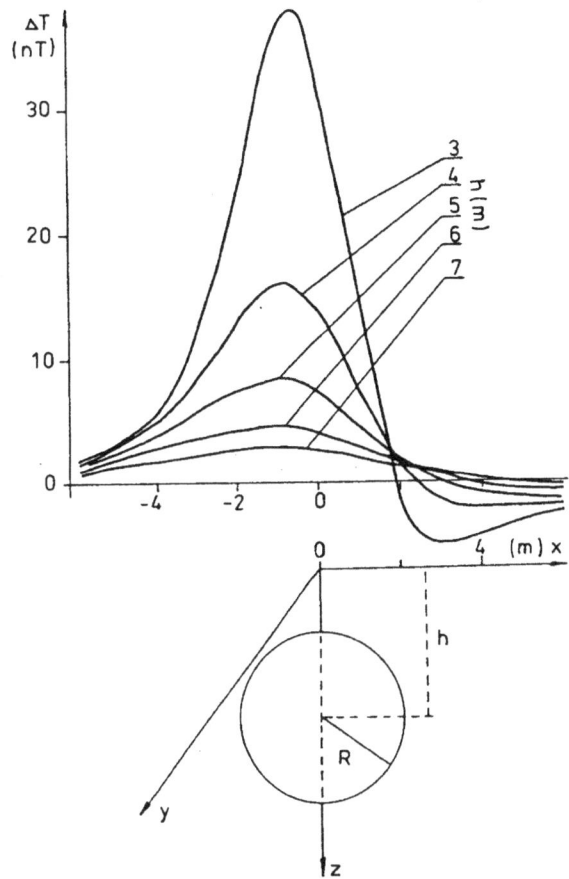

*Fig. 24. Magnetic effect of $\Delta T$ of a sphere with parameters $R=2\ m$, $\lambda = 60°$, $i= 65°$, $T_0 = 48,000\ nT$, æ $= 5.10^{-3}\ SI$: archaeological simulation - fireplace, furnaces, local point sources, etc.*

The calculation of the direct task for purposes of archaeogeophysical prospection, e.g. on a program calculator, can be implemented within three programs according to known formulas (Švancara 1984: 1985). These operations can be carried out at an arbitrary place of the profile normal to the structure of interest (Hašek et al. 1981).

1.  The SPHERE + CYLINDER program solves the direct magnetometric task for the model of a homogeneous sphere and a two-dimensional cylinder with a horizontal axis of rotation at the same time. The result is the calculation of the $\Delta T$ values on the profile with an equidistant step $\Delta x$ (see Fig. 24).

2.  The POLYGON program solves the direct task for two-dimensional objects set by the susceptibilities and coordinates of polygon points. For the description of the structure 41 (30) data registers are available, the program depositing susceptibility or the two coordinates of the polygon point $x_i$, $z_i$ into one program. Some more complex bodies are modelled by the set of various models of a prism (Hašek - Segeth - Vencálek 1990: 150-192) (see Figs. 25, 26).

3.  The STEP + PRISM program solves the direct task for the model of a vertical two-dimensional step and a two-dimensional prism (see Fig. 27).

Examples of models, their $\Delta T$ effects and archaeological objects that can be simulated by the model are given in Figs. 24 through 27. In each figure a set of curves is drawn for some parameter.

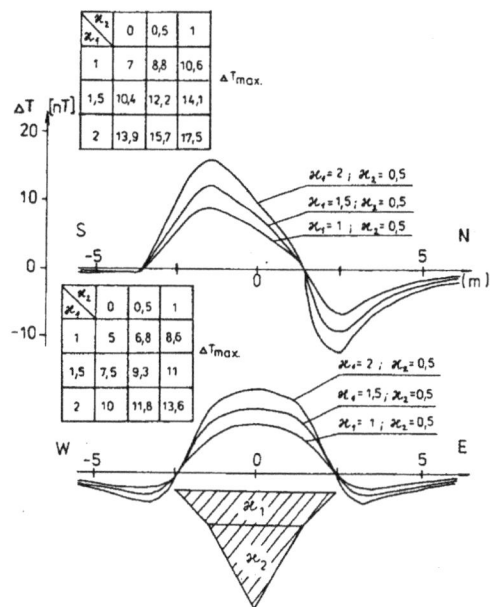

*Fig. 25. Magnetic effect $\Delta T$ of a trilateral prism of variable æ with the depth and parameters $h_1 = 0.5\ m$, $h_2 = 4.9\ m$, $\lambda = 90°$, $0°$, $i = 65°$, $T_0 = 48,000\ nT$ archaeological simulation - moat*

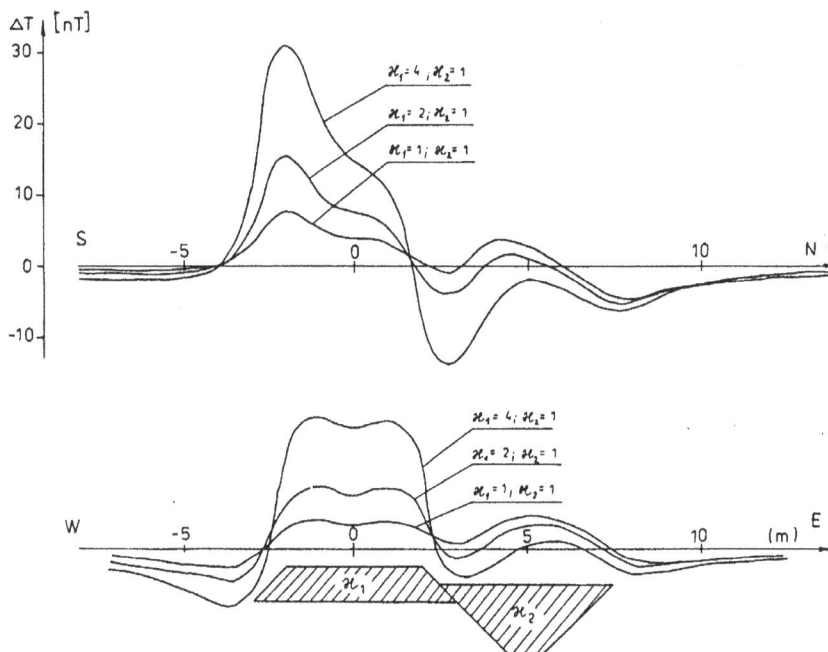

**Fig. 26.** *Magnetic effect ΔT of the system of two common prisms with variable æ₁ (æ₂ = const.) parameters $h_1 = 0.5\ m$, $h_1' = 1$ m, $h_2 = 1.5\ m$, $h_2' = 3\ m$, $\lambda = 90°$, $0°$, $i = 65°$, $T_0 = 48,000\ nT$ archaeological simulation - fortification system*

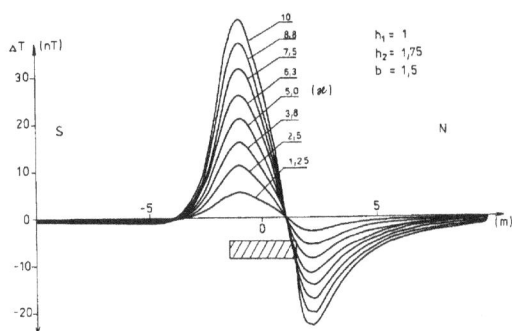

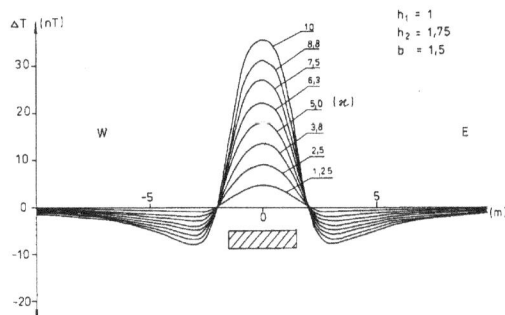

**Fig. 27.** *Magnetic effect ΔT of a two-dimensional prism with a variable æ, parameters $h_1 = 1\ m$, $h_2 = 1.75\ m$ $2b = 3\ m$, $\lambda = 90°$, $0°$, $i = 65°$, $T = 48,000\ nT$ archaeological simulation - recessed habitation objects, graves, production objects. etc.*

For the calculation of the direct task it is also possible to use the GAMA application software developed for purposes of gravimetric and magnetic modelling in solving the reverse task (Švancara-Halíř 1986) (see Chapter 3.3.2.2.). The mathematical model consists either of horizontal polygons, i.e. 2 1/ 2D model (Fig. 28) or of vertical polygons, i.e. a horizontal layered model (Fig. 29).

The polygons set individually delimit magnetically homogeneous parts of the modelled object and/or locality (Halíř - Hašek1989: 193-204).

In Figs. 30 through 32 models of prisms are submitted and their T effects in a profile that can simulate different types of fortifications, e.g. a moat with a depth variable magnetic susceptibility combined with the possibility of further archaeological objects, such as furnaces, palisade grooves, etc. From Fig. 30 it is evident that a minor object located within its body of the same susceptibility (æ₂) as the bottom layer of the moat filling, can be separated in a more complicated way from the calculated ΔT curve, the same as the palisade groove, although the two undistinguished objects can be assumed from the magnetic figure.

A more apparent expression at ΔT curve is in the smaller object with the susceptibility $æ_3 = 2.10^{-3}$ (SI) at the edge of the moat (Fig. 31). This fact was found e.g. in the space of the Neolithic elliptical structure near Vedrovice (district Znojmo), where at the outer edge of the "trough shaped" moat positions of furnaces with diameters of about 2 m were exposed (Hašek - Měřínský1991: 109-112).

A minor body located in the filling of the moat, such as a recent object or an object of $æ_3 = 2.10^{-3}$ SI is practically not projected onto the ΔT curve.

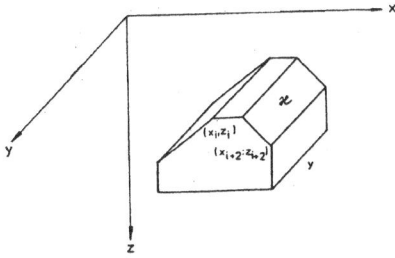

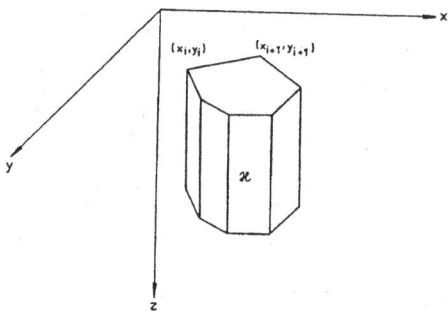

*Fig. 28 (Top). Geometry of arrangement of a 2.5D body, $x_i$ $z_i$ being polygon points delimiting the section of the body in the plane x,z, y is the body length*

*Fig. 29 (Bottom.) The geometry of arrangement of a 3D body, $x_i$ $y_i$ are polygon points delimiting the section of the body in the plane x y*

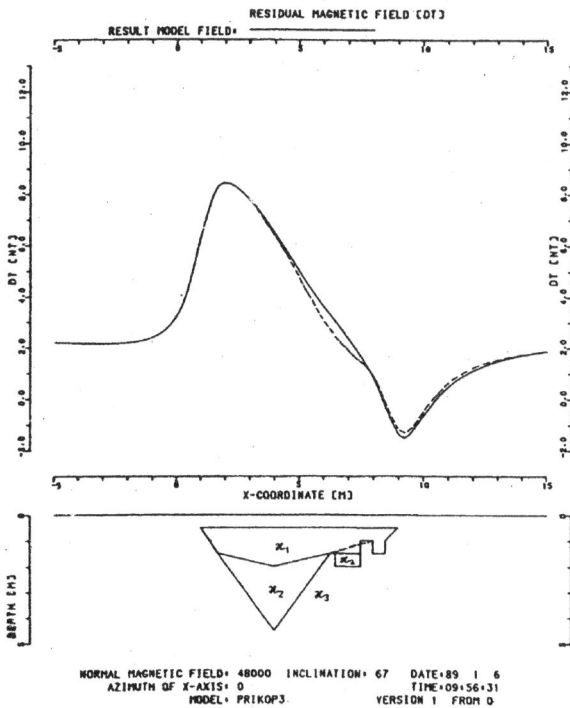

*Fig. 31. Magnetic effect of $\Delta T$ of common prisms with variable æ ($æ_1 = 1.4 . 10^{-3}$, $æ_2 = 0.9 . 10^{-3}$, $æ_3 = 2 . 10^{-3}$, $æ_4 = 0.2 . 10$ SI) simulating a moat with a palisade groove and two small objects, $\lambda = 90°$, $i = 65°$, $T_0 = 48,000$ nT.*

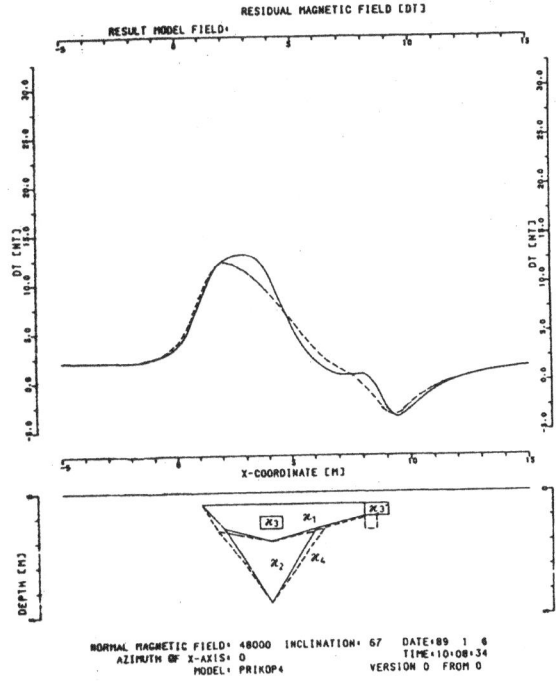

*Fig. 32. Magnetic effect of $\Delta T$ of a prism simulating a recessed object with a furnace ($æ_1 = 1 . 10^{-3}$, $æ_2 = 0.3 . 10^{-3}$, $æ_3 = 2 . 10^{-3}$ SI), $\lambda = 90°$, $i = 65°$, $T_0 = 48,000$ nT*

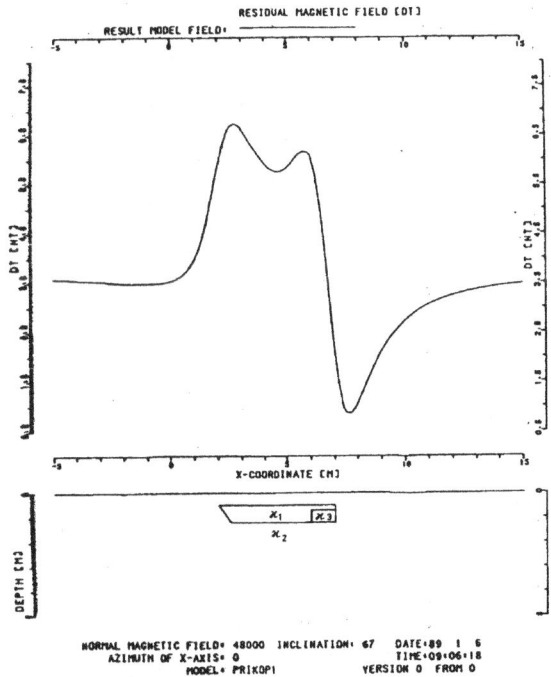

*Fig. 30. Magnetic effect of $\Delta T$ of common prisms with variable æ ($æ_1 = 1.4 . 10$, $æ_2 = 0.9 . 10$, $æ_3 = 0$ SI, simulating a moat with a palisade groove and a small object at its inner side, $\lambda = 90°$, $i = 65°$, $T_0 = 48,000$ nT*

In Fig.32 a possibility of separation of a recessed three-dimensional habitation object is shown, whose filling is a layer $æ_1 = 1.10^{-3}$ SI. A small object - a fireplace, a furnace, etc.( $æ_3 = 2.10^{-3}$ SI) can be easily located from the calculated curve $\Delta T$, unlike the similar case in Fig.30.

29

Besides the calculations of the magnetic ΔT effects of archaeological objects in the section (Figs.24 through 32) the above program can serve for modelling even the whole expression of the structure in the plane and for making a map of ΔT isanomales. The theoretical effect of ΔT of a Neolithic circular object (roundel) which was simulated by 12 general prisms ($h_1 = 0.5$ m, $h_2 = 4.5$ m) interrupted in four places by entrances to the object is given in Fig. 33. The resulting map of ΔT isanomales corresponds to the results of processed field data (Hašek - Měřínský1989: 45-53).

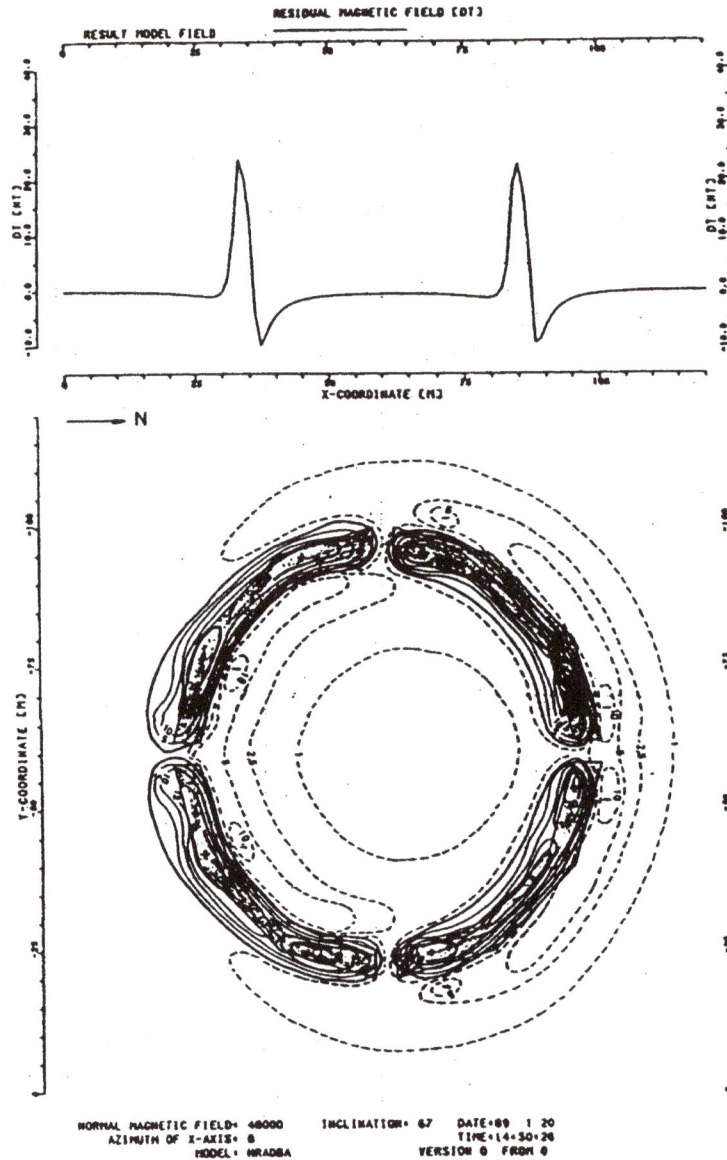

*Fig. 33.* A type of model simulating a simple Neolithic circular formation

## Modelling Electric Fields

### I. Resistance Profiling

The issues of modelling spherical and cylindrical nonconductive as well as conductive bodies were dealt with by Anders (1971: 5-9, 1972: 12-17); Brizzolari (1975: 1-12), Carabelli (1967: 9-21), Habberjam (1969: 780-785), Laksahmann (1963: 9-15); Parasnis (1965: 3-69); Töepfer (1969: 791-220), Withe (1970) and a number of other authors. All these papers of prevailingly theoretical character have the disadvantage that they assume only a simple and approximate solution of the task, such as a disturbing body in a homogeneous electric field, the possibilities of the Wenner and Schlumberger arrangement of electrodes for purposes of finding the above inhomogeneities, etc.

Mathematical modelling of nonconductive cylindrical bodies by means of different methods of resistance profiling was dealt with only by Lösch-Militzer-Rössler (1979: 53-126); Militzer-Rössler-Lösch (1979, 640-652). In these papers the general calculation of the curves of the apparent specific resistance is deduced starting from the solution of the Laplace differential equation. According to the calculation of the above authors it is possible to submit also some $\rho_a/\rho_1$ curves above the cylindrical body and graphs of "anomalous effects" (A.E.) for a further analysis in dependence on the individual geometries of electrodes and/or depths of anomalous bodies (Figs. 34 through 36).

From the model curves (Figs. 34 and 35) compiled for the purposes of the archaeogeophysical prospection it follows that the Wenner profiling has a great dependence of A.E. on the distance of electrodes, such as the Schlumberger electrode arrangement. With the same distance of AB and/or AO the geometry with MN < 1/3 AB, and/or MN < 2/3 AO yields more intense anomalies in the middle part of the $\rho_a/\rho_1$ curve. The method of the middle gradient stresses the anomalies of the apparent specific resistance. In dipole profiling, with unsuitable distance of electrodes, it is possible to measure more complicated anomalies $\rho_a$ which complicate the interpretation of the measured data. With the three-electrode (potential, gradient) configuration it is possible to use either the classical method of resistance profiling (one of the branches of the KRP curve, i.e. AMN or MNB and/or both branches of these curves), or a more effective way, i.e. the differential arrangement (AMN$_{dif}$). The profile curves AMN$_{dif}$ have inflexion points above the anomalous body and increase the value of A.E. more than the discussed arrangements. By decreasing MN at the same AO the magnitude of A.E. is in parallel increased. In Fig.36 are shown curves of the differential potential (MAN) and the gradient arrangement of electrodes (AMNA') together with A.E. in the presence of a two-dimensional nonconductive cylindrical body (according to Militzer - Rössler - Lösch 1979: 640-652).

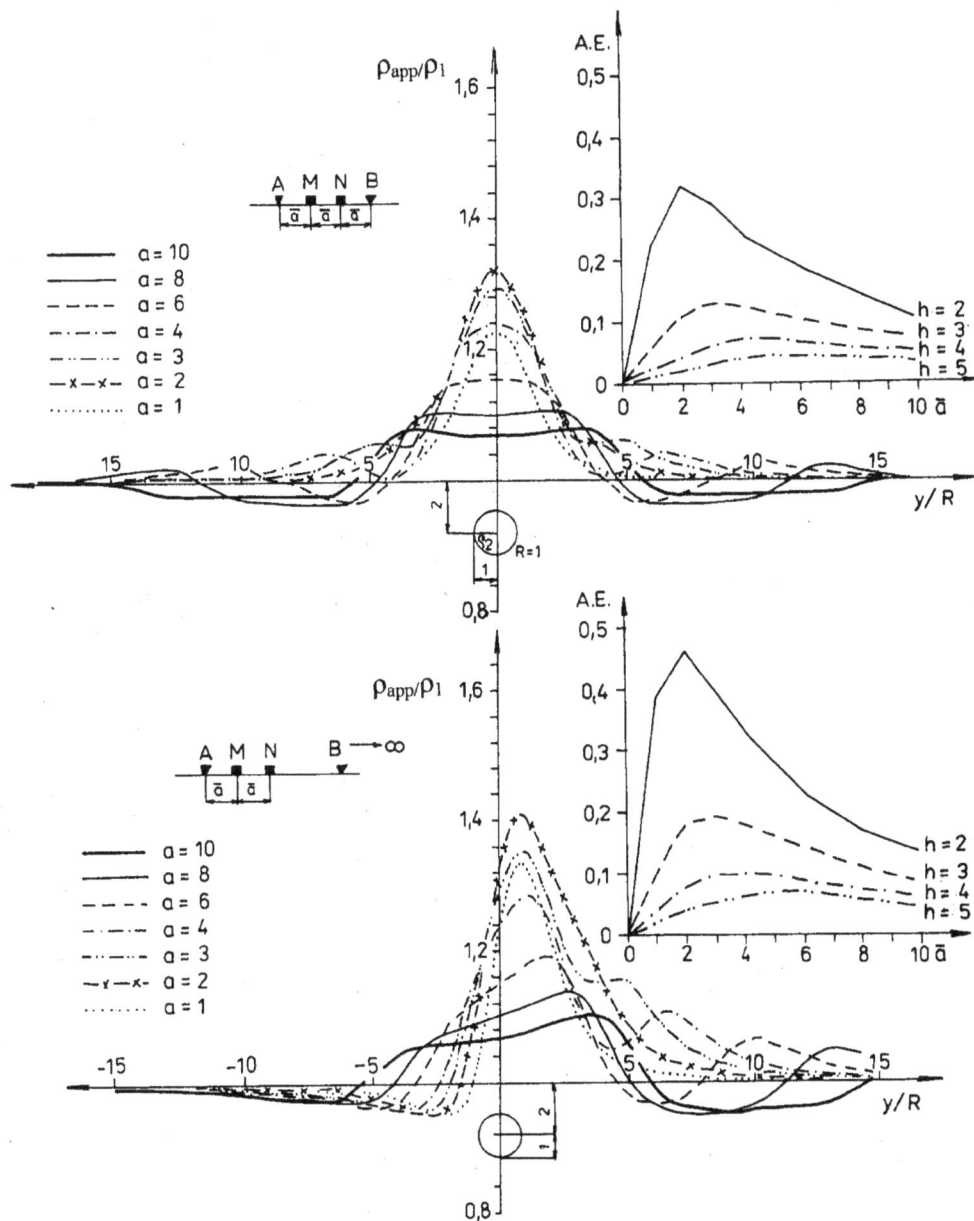

*Fig. 34. Profile curves of the potential electrode arrangement (Wenner, three-electrode) and A.E. graphs for a nonconductive cylindrical body - masonry, cavities, fortifications, etc.*

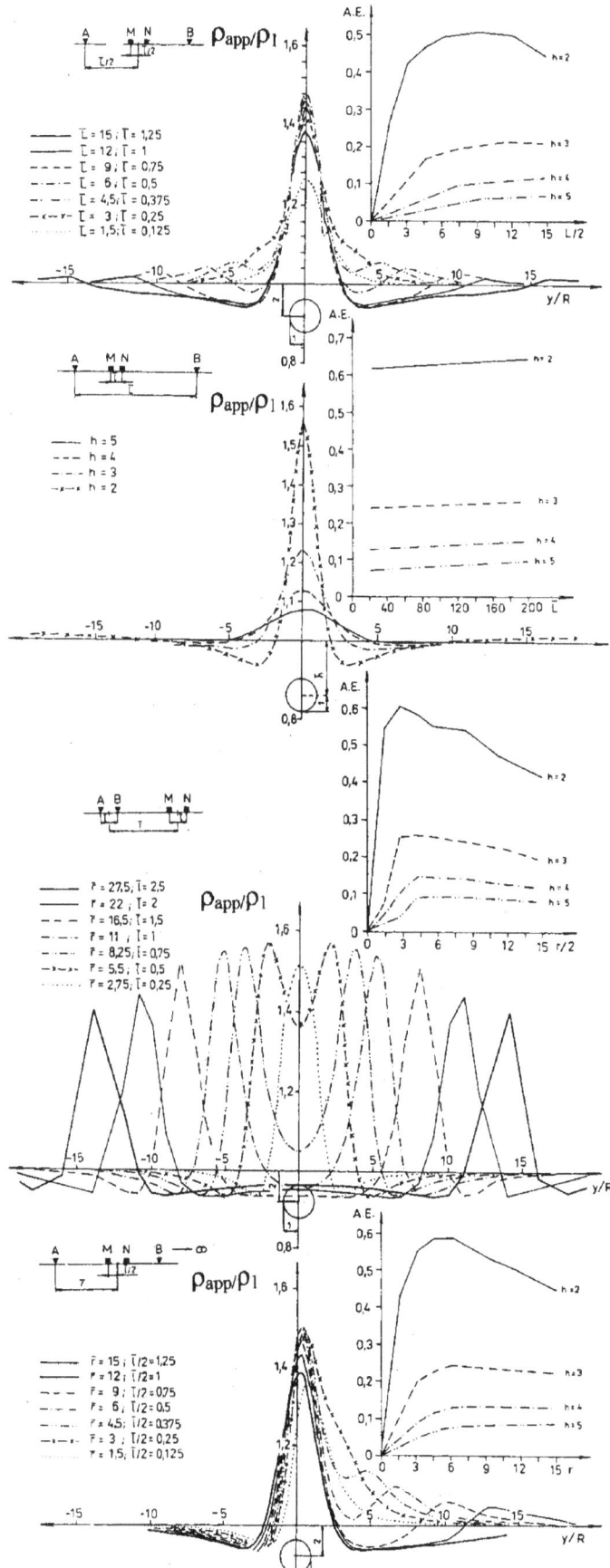

**Fig. 35.** *Profile curves of gradient electrode arrangements (Schlumberger, central gradient, dipole axial, three electrode) and A.E. graphs for a nonconductive cylindrical body*

The two geometries of electrodes are suitable for monitoring objects situated in a homogeneous environment. However, they do not allow the calculation of the apparent specific resistance. They belong to the so-called methods of "pure anomaly" as stated in Chapter 3.1.2.1. According to Gupta-Bhattacharya (1963: 608-616) the current density has a maximum for the AMNA' arrangement at the depth $h/l = 0.71$ (where l is the half distance of the current electrodes). The maximum is denoted as the "focusing" the current in a homogeneous environment, and thus these arrangements are denoted as focused. With $h/l>2$ the current density of the

focused arrangement is greater than in the four-electrode one. As the depth of the maximum current density can be controlled by the choice of a suitable electrode configuration, also the effect of the searched inhomogeneity can thus be obtained.

From Fig. 36 it is evident that the differential (potential, gradient) arrangements theoretically better locate cylindrical bodies at different depths below the Earth's surface than the hitherto currently used electrode geometries.

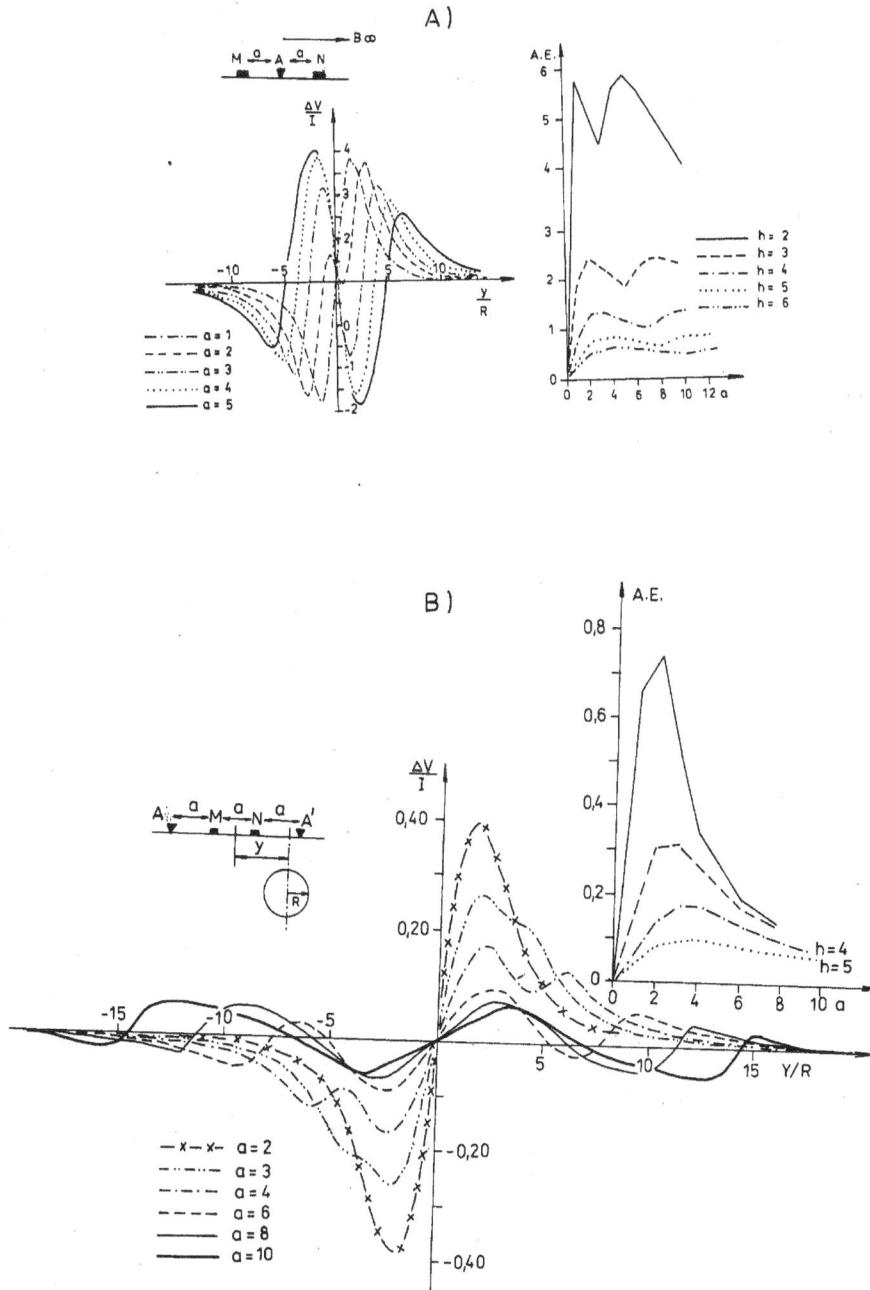

Fig. 36. Profile curves of a differential potential and a gradient electrode arrangement and A.E. graphs for a nonconductive cylindrical body

Theoretical and practical problems of modelling vertical resistant bodies at different electrode arrangements were dealt with by Kumar (1973: 560-578; 1973a: 615-625), Jain (1974: 445-457), Vešev (1980), and others.

In Slovakia the direct task for a non-conductive plate (masonry, a defence line - vallum and a trilateral prism-moat) was solved by Tirpák (1984); Hvožďara - Tirpák (1987: 165-188). The methodology of the papers is based on a numerical and physical modelling of electric fields due to a line or point current source for two-dimensional resistance inhomogeneities. For the calculation of the potential field of stationary current in a semispace and the inhomogeneity of the above type it uses the method of integral equations

elaborated by Hvožďara - Schlosser (1985: 35-49). The calculation of electric potentials is solved by means of Green's formula, i.e. the summary effect of dipoles induced to the surface of the disturbing body is considered.

Numerical and physical modelling of the studied inhomogeneities was applied for different electrode arrangements (AMNB, AMBN, ABMN, AMNB, AMNB) and the depth of anomalous bodies. It is possible to give some practical examples, the course of the curve $\rho_a/\rho_1$ above the nonconductive plate (Fig. 37) with a trilateral prism (Fig. 38) at the Wenner and the Schlumberger profilings which are most frequently used for solving the above types of tasks.

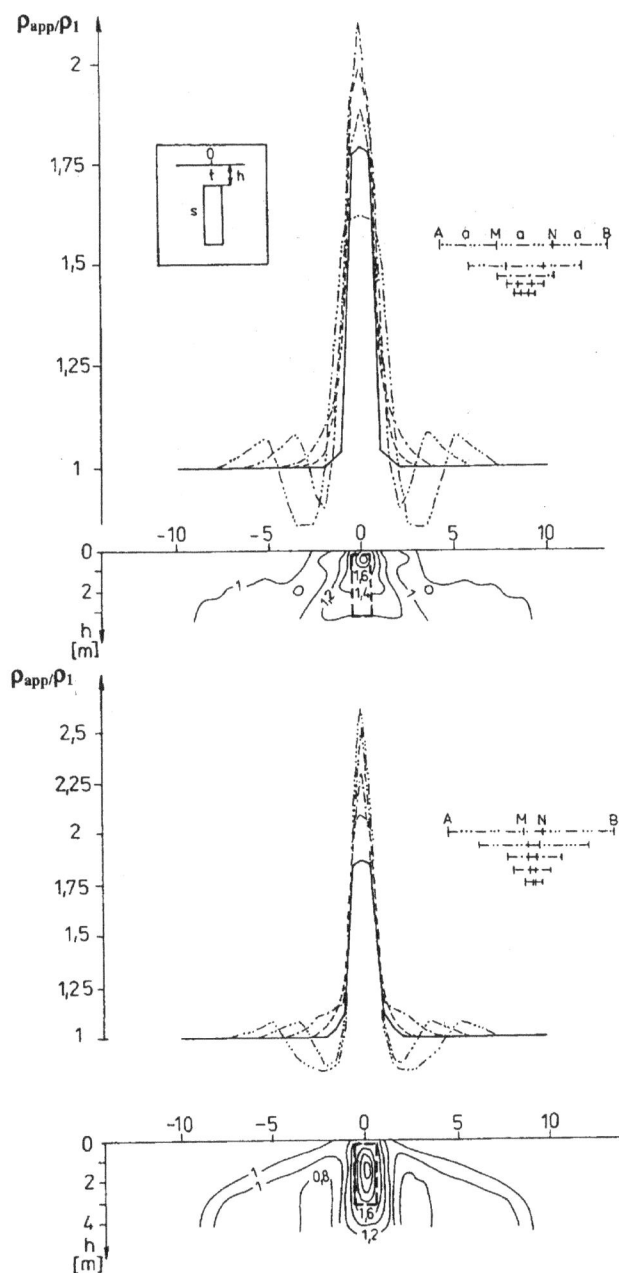

*Fig. 37. Profile curves of the Wenner and the Schlumberger electrode arrangements and an isoohmic section for a nonconductive plate shaped body - masonry, fortifications, tombs, etc.*

From Fig. 37 it follows that for the case when the distance between the individual electrodes and t (the thickness of the body) theoretical $\rho_a/\rho_1$ curves above the plate have a simple course characterized by one maximum. The width of the anomaly in the middle of the maximum corresponds to the double thickness of the studied body.

With a>t, besides the maximum above the centre of the body there also appear accompanying maximums on both sides. The widths of the anomaly in the half size of the maximum of the electrode arrangement exhibit a simple maximum above the nonconductive plate at L/3 < t. At L/3 > t also accompanying minimums appear on the curves besides the maximum, as in the preceding case. The width of the anomaly in the half value of the maximum corresponds for all cases to the double thickness of the studied body.

For the case of the trilateral prism the theoretical $\rho_a/\rho_1$ curves in the Wenner electrode arrangement (Fig. 38) are characterized for a<t by one maximum, where the width of the anomaly at the half of the maximum corresponds to the length of the t-side of the prism. For the case a ≥ t the curves have two maximums. Their mutual distance indicates the approximate dimension of t.

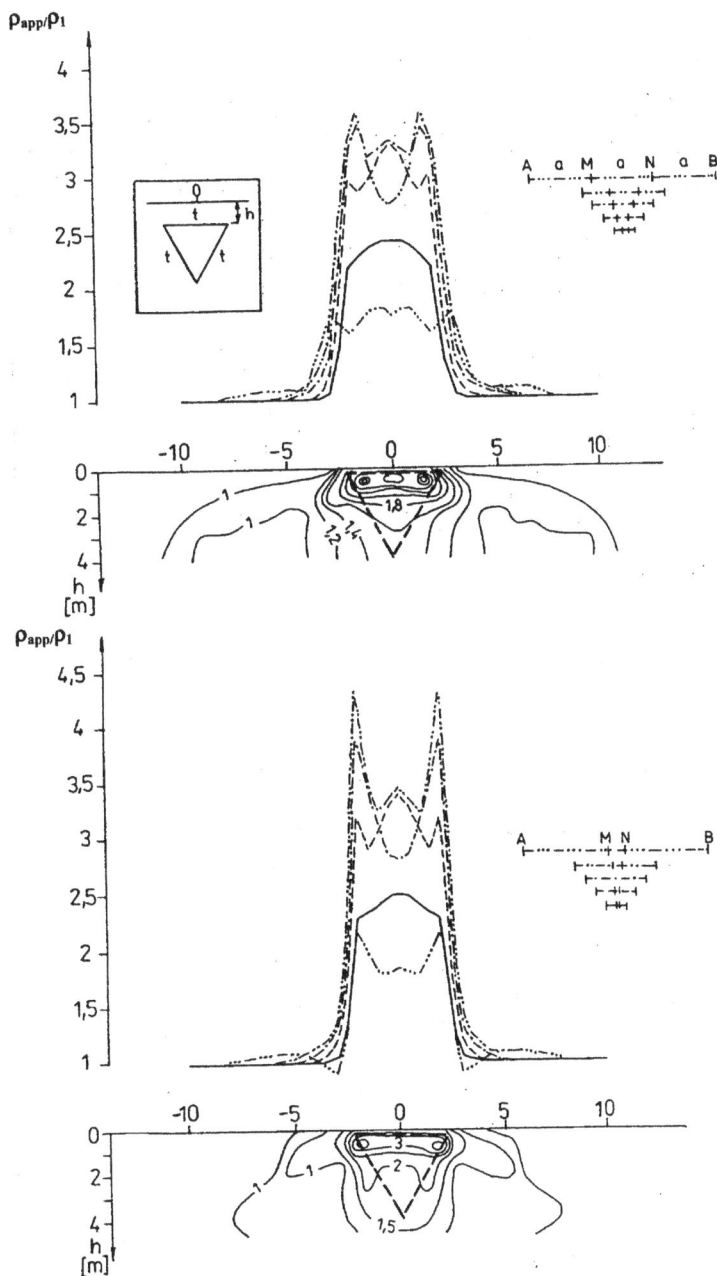

*Fig. 38. Profile curves of the Wenner and the Schlumberger electrode arrangements and an isoohmic section for a nonconductive trilateral prism - a moat*

35

Similar properties are also exhibited by theoretical curves of the Schlumberger arrangement which have a simple maximum at $1/3 < t$. At $L/3 > t$ the $\rho_a/\rho_1$ curves are characterized by two maximums whose mutual distance is the length of the t-side. For the case when $L/3 \gg t$, besides the maximum above the body there also appear accompanying minima along both sides. Gajdoš - Tirpák (1989: 262-269) modelled also the dependence of the $\rho_a/\rho_1$ curves for a horizontal prism and a vertical one in the Wenner, the Schlumberger and the dipole arrangements at

a) the horizontal width of an anomalous body (a) and the distances of the electrode system (L).,

b) the vertical dimension of the body (b),

c) the depth of the anomalous body.

From the constructed model profile curves of $\rho_a/\rho_1$ it can be deduced that

1) the size of the amplitude of the anomalous response increases with increasing horizontal dimension of the body under the assumption that $L > a$ in all electrode arrangements. The maximum anomalous response is exhibited by curves of the dipole profiling, unlike the Wenner and the Schlumberger profilings which have approximately the same amplitude of $\rho_a/\rho_1$,

2) the vertical dimension of body (b) in case that $b > 1/6\ L$ has no substantial effect on the shape of the anomalous response. With increasing b the size of the amplitude of the anomalous response increases in the Wenner and the Schlumberger electrode configurations. In the dipole profiling the change in the vertical dimension of the body does not practically change the size of the amplitude of the anomalous response.

3) the depth of the anomalous body affects the size of the amplitude of the anomalous response. It is a greater effect than that in the vertical dimension of the body.

## II. Vertical Electric Sounding

The modelling in VES can be implemented by the calculation on PC, or even by a graphical construction of three theoretical or multilayered curves of VES for the selected model of horizontally layered environment of the studied archaeological structure (thicknesses and specific resistances of the individual geoelectric layers of the section - $h_i\ \rho_i$), according to the procedures included in different monographs and publications, such as Gruntorád-Karous (1972); Mareš et al. (1979: 284-293); Hašek(1972: 333-336); Hašek et al. (1984a); Chyba (1981), and others. In archaeological prospection it has not yet had a broad application.

## III. Dipole Electromagnetic Profiling

In mathematical modelling of the effects of simple geometrical bodies, such as prisms located in the environment with different conductivities, for the employment of apparatus (KD-1, EM-38) one starts from the functional dependences characterizing the relative contributions (responses) of horizontal layers to the secondary magnetic field or to the measured apparent conductivity ($\sigma_a \equiv \sigma_{DEMP}$).

The functions $F_V(z)$ and $F_H(z)$ (McNeil 1980: 5-14) which are in the form

$$F_V(z) = 4z/(4z^2+1)^{3/2} \qquad (27)$$

$$F_H(z) = 2 - 4z/(4z^2+1)^{3/2} \qquad (28)$$

represent the dependence of the relative contribution to the secondary magnetic field of the unit layer at the standardized depth ($z = z_i/r$, $z_i = h_1+h_2+\ldots\ldots h_i$) of its deposition for ZZ and YY polarizations. For calculating the response of electric conductivity of the whole measured semispace (Ro) is there the following expression

$$Ro = \int_0^\infty F(z)dz \qquad (29)$$

By the adaptation in formula (29), i.e. substituting for the lower limit the general standardized depth z it is possible to write for $R^V_H(z)$

$$R_H^V(z) = \int_z^\infty F_V(z)dz \qquad (30)$$

By substituting expressions (27), (28) into (30) and by integration

$$R_V(z) = 1/(4z^2+1)^{1/2} \qquad (31)$$

$$R_H(z) = (4z^2+1)^{1/2} - 2z \qquad (32)$$

By means of formulas (31) and (32) it is possible to simply find the conductivity above a multilayer horizontal environment on the assumption that $r/\delta \ll 1$. The $R^V_H(z)$ function for apparatus EM-31 and EM-38 are given in Fig. 39. By multiplying them for different z by the value 100 % (or by conversion to a different height of the apparatus) we obtain the percentual response of the underlier below the chosen depth on the assumption that the conductivity does not change with depth.

By means of formulas (31) and (32) it is simple to calculate $\sigma_a$ above two- or multilayer environments.

Hence the formulas for a two-layer case (McNeil 1980: 5-14):

a) the contribution of conductivity from the first layer with parameters $z_1$

$$\sigma_a = \sigma_1 (1 - R_V(z_1)) \qquad (33)$$

b) the total contribution of the conductivity of the second layer

$$\sigma_a = \sigma_2 R_V(z_1) \qquad (34)$$

or measured $\sigma_a$ will be in the form

$$\sigma_a = \sigma_1(1 - R_V(z_1)) + \sigma_2 R_V(z_1) \qquad (35)$$

If one starts from formulas (33 through 35), then for a three-layer environment:

$$\sigma_a = \sigma_1(1 - R_V(\vartheta z_1) + \sigma_2(R_V(z_1) - R_V(z_2)) + \sigma_3 R_V(z_2) \qquad (36)$$

The general recurrent relation for the n-layer case

$$\sigma_a = \sum_{K=1}^{n} \sigma_K ((Rv(z_{K-1}) - Rv(z_K)) \qquad (37)$$

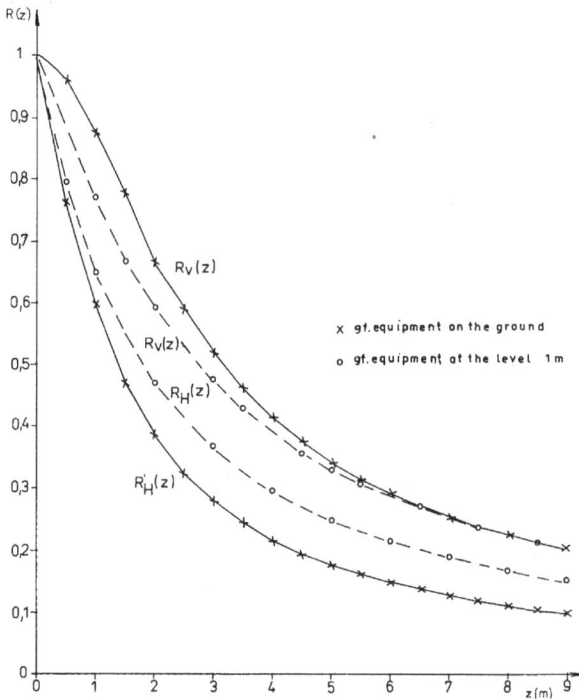

*Fig. 39. Relative response to a secondary magnetic field of the whole environment located below the standardized depth level for ZZ and YY-polarizations in apparatus KD-1 and EM-38*

For the conversion to the apparent specific resistance formula (3) is used. On several simple cases it is possible to show some practical possibilities of modelling in the DEMP method by means of equations (35) and (36).

By means of the adapted formula (35) it is possible to calculate $\rho_a/\rho_1$ above a trilateral nonconductive prism for different heights and positions of apparatus KD-1, EM-31 and EM-38 simulating an ideal case of the effect of a moat (Fig. 40). Its anomalous response at the chosen parameters of the environment is relatively marked in the two instruments both at ZZ and at YY polarization. In measuring with apparatus EM-38 located on the ground the amplitude of the anomaly in the two polarizations above the mentioned structure is the highest (unlike KD-1 at the height of 1 m), even though at YY its size is lower by about 4% than at ZZ.

According to formula (36) was made the calculation of $\rho_a/\rho_1$ over various bodies, particularly prisms, simulating in ideal cases recessed habitation and production objects or fortifications (Fig. 41), masonry and graves (Fig. 42). The minimum limit for finding them was chosen in a similar way as in the DC resistance methods the magnitude 0.1 A.E. which corresponds to the increase or decrease $\rho_a$ in the

anomalous region about 10% of the "undisturbed" environment (Hašek et al. 1981).

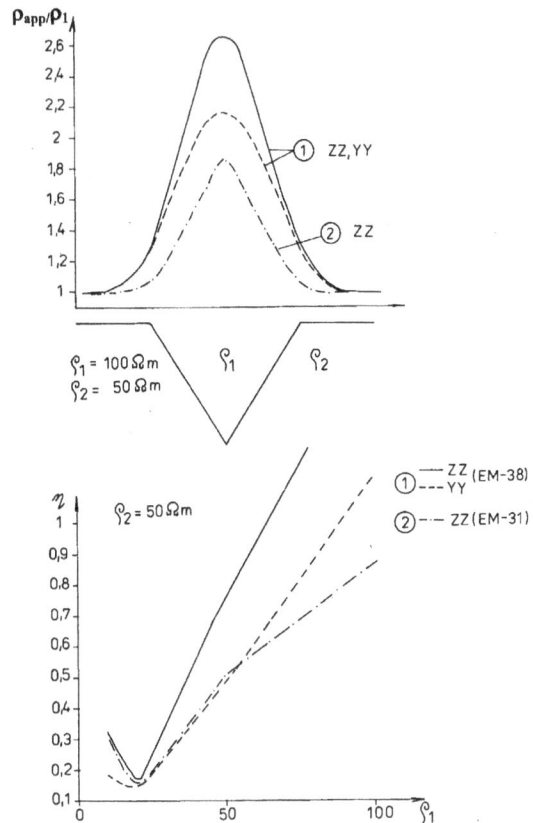

*Fig. 40. Profile curves of $\rho_a/\rho_1$ above a nonconductive trilateral prism with parameters $\rho_1 = 100$ ohmm, $\rho_2 = 50$ ohmm, $z_1 = 0.5-4.5$ m, simulating a moat*

Under this assumption it holds for the above examples that
a) in shallow bodies with the depth to about 1.5 m (Fig. 41) and with the variables $\rho_1$ and $\rho_2$ it is possible to locate these models by means of the apparatus EM-38 for different $\rho_2$ of the order of up to 40 ohmm. At $\rho_a > 40$ ohmm it is impossible to follow the object. The greater the ratio between $\rho_2/\rho_1$, such as 1/10, the higher is also $\eta$ and vice versa, at YY polarization $\eta$ drops (size of anomaly) for small $\rho_1$ of the order by 1/3 to 1/4 $\eta$ against the ZZ polarization.
b) when finding the masonry foundations by means of the two above apparatus (Fig. 42) its use can be expected under the assumption that $z_1$ does not exceed the depth of 2 to 2.5 m at the parameters $\rho_1 = 500$, 1000 ohmm and $\rho_2 = 100$ ohmm, in case $\rho_1 = 100$ ohmm, $\rho_2 = 1000$ ohmm, $\rho_3 = 200$ ohmm, then from EM-31 (height 1 m) these objects can be identified up to $z_1 = 2$ m and with EM-38 up to about 1-1.5 m.

The applications of electromagnetic methods in modelling cylindrical bodies has so far not been paid such attention as they would deserve. There is a number of papers concerning only the theoretical solution of some problems of the given task. Faldus et al. (1963: 372-393) dealt with model calculations and with the measuring of the effect of a spherical cavity on the electromagnetic field. Thus he links up with the papers by Wait (1954: 20, 1955: 630) which

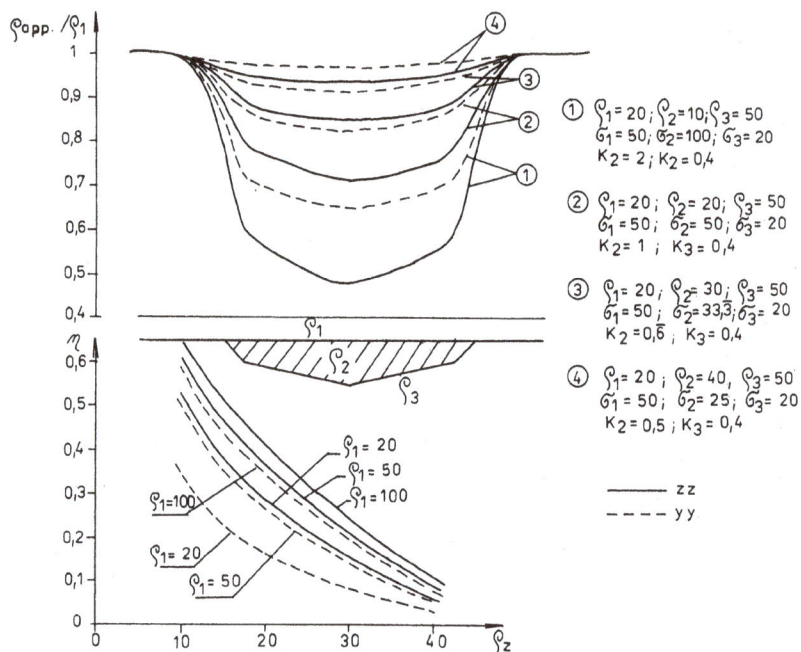

**Fig. 41.** *Profile curves of $\rho_a/\rho_1$ and graphs above a plate shaped body for different values of $\rho_1 = 20, 50, 100$ ohmm, $\rho_2 = 10, 20, 30, 40$ ohmm, $\rho_3 = 50$ ohmm, orientation of apparatus -ZZ, YY (EM-38)*

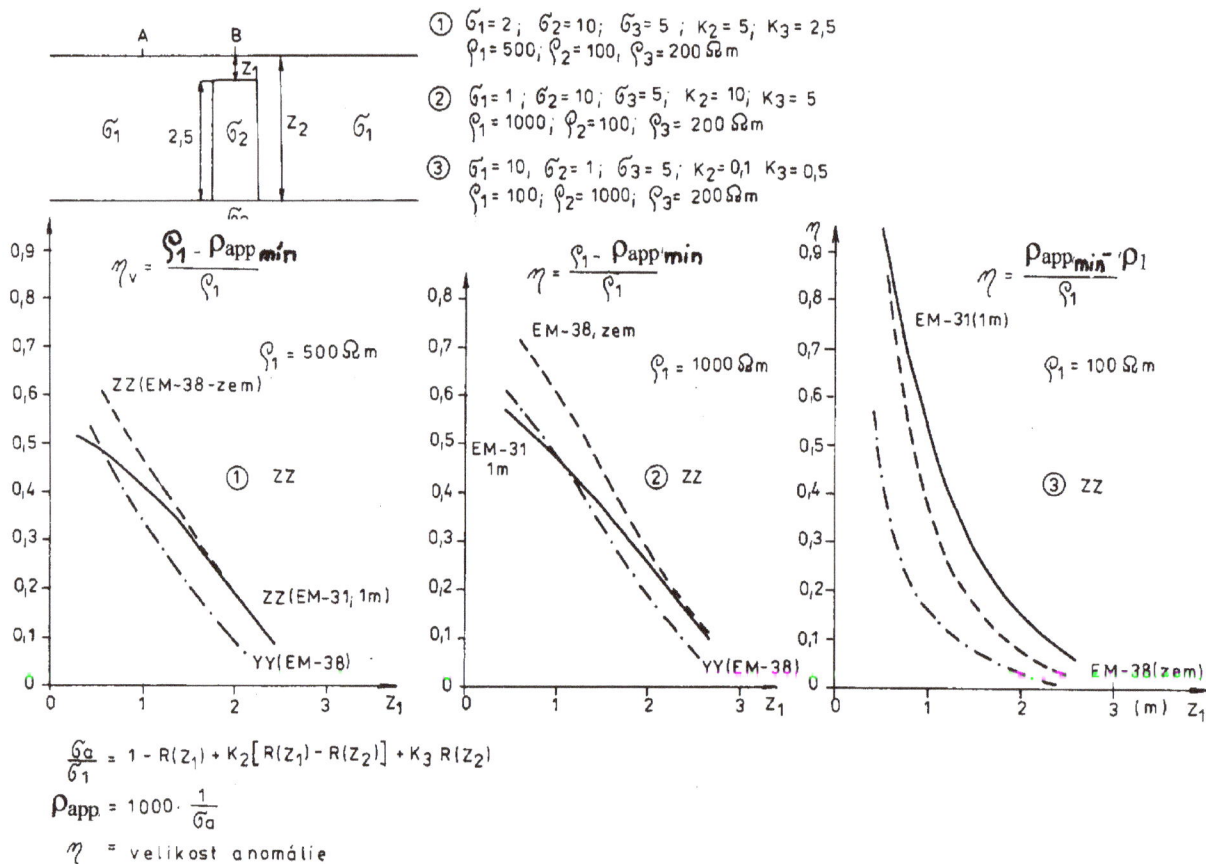

$$\frac{\sigma_a}{\sigma_1} = 1 - R(z_1) + K_2\left[R(z_1) - R(z_2)\right] + K_3 R(z_2)$$

$$\rho_{app} = 1000 \cdot \frac{1}{\sigma_a}$$

$$\eta = \text{velikost anomálie}$$

**Fig. 42.** *Size of anomaly for apparatus KD-1 and EM-38 above a body ($\rho_1 = 100, 500, 1000$ ohmm, $\rho_2 = 100, 1000$ ohmm, $\rho_3 = 200$ ohmm) simulating masonry, recessed objects, graves, etc.*

studied the effect of the electromagnetic relation transmitter-inhomogeneity-receiver. Model studies with a conductive sphere in the electromagnetic field of the vertical magnetic dipole were carried out by Molochnov and Balobayev (1958: 80-89). The authors stress a perceptible effect of the inhomogeneity on the phase angle between the vertical and the horizontal components of the magnetic field, i.e. the effect known from the theory of wave. The mathematical solution of the given issue was further dealt with by Latka (1966: 512-517) and Jones-Pascal (1971). Buchhein (1952) described theoretically the effect of inhomogeneities of near-surface layers on the electromagnetic field. His theory approaches the actual geoelectric conditions. Theoretical and practical processing of electromagnetic fields in relation to the employed frequencies and the distance transmitter-receiver was solved by Bláha et al. (1973); Bláha-Chyba (1978). The possibilities of detectability in small cylindrical bodies by the DEMP method has recently been dealt with by Záhora (Hašek et al. 1988). From the results of numerical calculations there follows a relative complication of the solution of this task given both by the dimensions of the searched for inhomogeneities and their depth as well as by the measuring system.

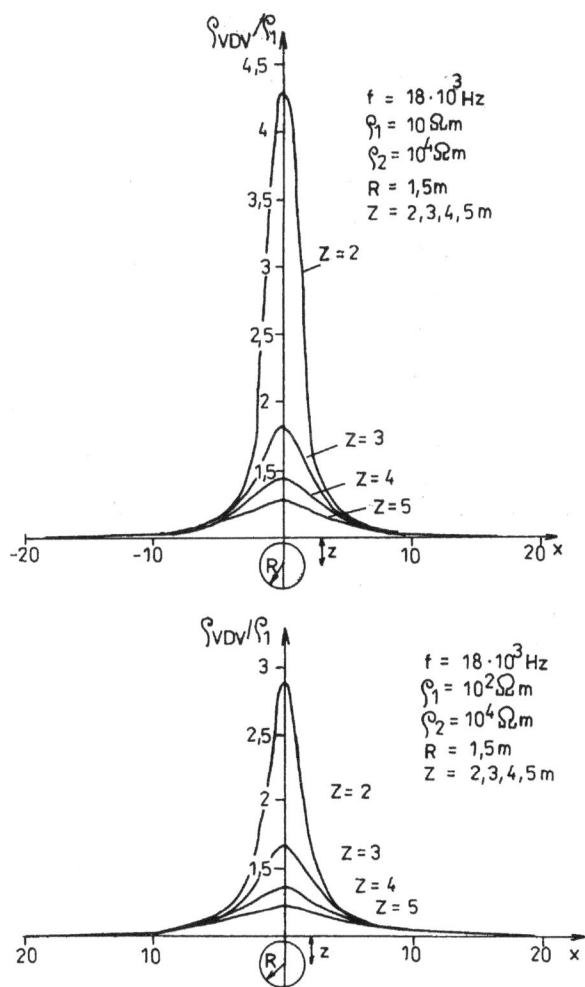

*Fig. 43. Profile curves $\rho^{VLW}_H/\rho_1$ above a nonconductive cylindrical body simulating a cavity, masonry, etc.*

## IV. The Resistance Version of the VLW Method

Mathematical modelling of simple bodies (cylinder, sphere, plate, contact) for the VLW-R method was dealt with by Bláha - Chyba (1978) and others. The basis of the theory is the development of the plane wave into a series of the Bessel or the Mathien functions. The model of a conductive body is a circular cylinder for which a number of rules have been set, according to which it is possible to estimate the behaviour of the individual parameters of the field. Theoretical curves calculated for the studied body can also be used for the quantitative estimate of effects of more complicated formations.

By the analysis of the calculated theoretical $\rho_H^{VLW}/\rho_1$ curves above the sphere and the conductive cylinder it was found that the anomalies above the sphere are of a lower order than those above bodies of cylindrical shape. The determination of 3D bodies by the resistance variant of the VLW method is thus possible only under particularly favourable geological conditions.

The calculation of theoretical curves for a nonconductive cylindrical body in a semispace was made by Hašek et al. (1981). For the solution of the task he used an adapted program compiled by Chyba (Bláha - Chyba 1978) for a conductive cylinder. A practical example from the calculated $\rho_H^{VLW}/\rho_1$ curves is given in Fig. 43.

From the solution it followed that in the case of the existence of the studied environment without great thicknesses of conductive deposits nonconductive objects of cylindrical shape of diameter 3 - 4 m could be followed under favourable conditions to the depths of about 5 - 7 m (Hašek - Měřínský 1991: 65).

### 3.3.2.2. Solving the Reverse Task

#### The Magnetometric Method

Solving the reverse task can be carried out by means of the GAMA application software (Švancara-Halíř 1986), in which only "sufficiently" intense $\Delta T$ anomalies are interpreted. As a "sufficiently" intense can be considered such an anomaly whose intensity is in at least 3 points at least 3 times higher than the potency of the noise, i.e. the credibility is higher than 95%.

The input values for the computer are the measured magnetic data, the intensity of the geomagnetic field, inclination, profile bearing, susceptibility of the object, coordinates of its cusps and the distances of lateral sides 2 1/2D of bodies with respect to the position of the calculated profile. Of great importance is also the possibility to calculate the effects of the model with a marked share of the remanent magnetic polarization.

The effect of the model (Rasmussen - Pederson 1979: 749-760) is compared with the measured data and in case of discrepancy the result is optimized by means of gradient methods (conjugation by gradients - Fletcher - Powell 1963: 149-154; Marquardt 1963: 431-441; Meyer 1970) with the simulation of the limitation respecting the primary information about the studied locality. The variables can be cusps of the polygon, the susceptibility of bodies which can be associated, i.e. a group of variables can change by the same value, in that way it is possible to achieve a change in the model, such as the shift of the individual polygons and/or interfaces with preserving the shape, etc. Despite that, even at a good approximation it is necessary to realize the limitation of the model consisting in the shape of the bodies and, above all, in the assumption of a homogeneous distribution of susceptibilities in the objects of interest (Halíř - Hašek1989: 193-204).

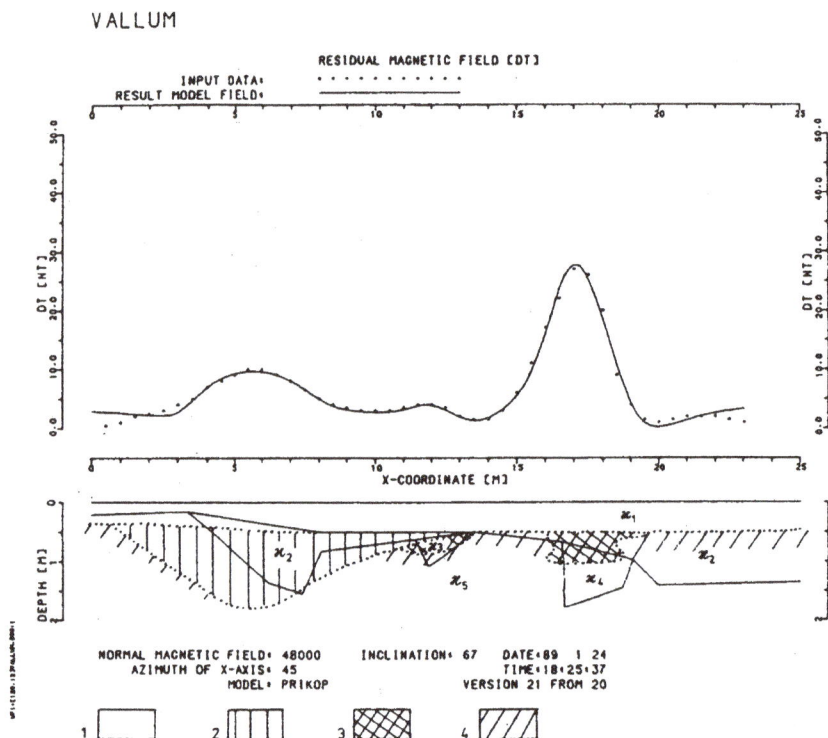

*Fig. 44. A comparison of the resulting model ($æ_1 = 1.73 . 10^{-3}$, $æ_2 = 1.49 . 10^{-3}$, $æ_3 = 1.51 . 10^{-3}$, $æ_4 = 7.35 . 10^{-3}$, $æ_5 = 0.24 . 10^{-3}$ SI) with the reality at the interpretation profile of the Slavonic ringwall Pobedim, district Trenčín (Slovakia) 1- topsoil, 2- moat filling, 3- burned wall destruction of the moat body, 4- sterile environment*

As an example, in Fig. 44 there is a solution of the reverse task in the section when monitoring the fortification system consisting of a burned body of the defence line - vallum and a moat at the Slavonic ringwall Pobedim (district Trenčín) (Hašek - Ludikovský- Obr 1979: 45-48).

The resulting physico-archaeological model matches with a remarkable precision the excavated structure (denoted by dotting), in some sectors, however, the result of the interpretation somewhat differs from reality. This can be explained by the selected input model of the fortification system, when e.g. for the moat and the burned layer of clays from the body of the defence line - vallum a susceptibility invariable with depth was used and a marked share of remanent magnetic polarization was not considered, as a result of which there can even occur a slight shift of sizes and depths of the separated anomalous bodies.

The calculation and compilation of the resulting areal physico-archaeological model in the space of the metallurgical centre dating back to the younger Roman time at Sudice (district Blansko) proved a very good coherence between the interpreted anomalous body and the reality verified by archaeological investigation (Fig. 45).

Its assumed size covers with a relatively great accuracy the space in which 72 hearths of metallurgical facilities were found with the diameter of 40 - 45 cm, irregularly distributed on an area of about 64 m$^2$ (Halíř - Hašek1989: 193-204; Hašek - Měřínský 1991: 152-157).

For finding the characteristic of the sources of linear anomalies $\Delta T$ it is possible to use with success also the method of the deconvolution of the profile curves (Karousová 1979). The application of this method allows us to find the distribution of fictive bodies (such as cylinders, thin plates) at the set depth below the measured profile and determine their dimensions. Although the shapes are not identical with the real ones, the location of these bodies gives a clearer idea about the position and size of the object searched for than the originally measured $\Delta T$ curves. By the above method a matrix of physical parameters of a cylinder is obtained (M = J.S., where J is induced magnetization, S the area of the perpendicular section of the cylinder), or that of a thin vertical plate (M = 2J.2b; 2b being the thickness of the plate) whose columns correspond with the individual profiles.

In the compiled program, for the sake of higher accuracy, nine-component filters were chosen for the two cases. For different directions of profiles they are in the form

$$M^{VD}(x) = \sum_{i=-4}^{4} G_i \, \Delta T(x-ix) \qquad (38)$$

where

$G_i$ are the filter coefficients changed at different directions of the profiles.

An example of processing the deconvolution curves $M^V(x)$ in delimiting the size and relative depth of a Neolithic circular structure from the locality of Němčičky (district Znojmo) is given in Fig. 46 (Hašek - Segeth - Vencálek 1990: 156-192; Hašek - Měřínský 1991: 106-107).

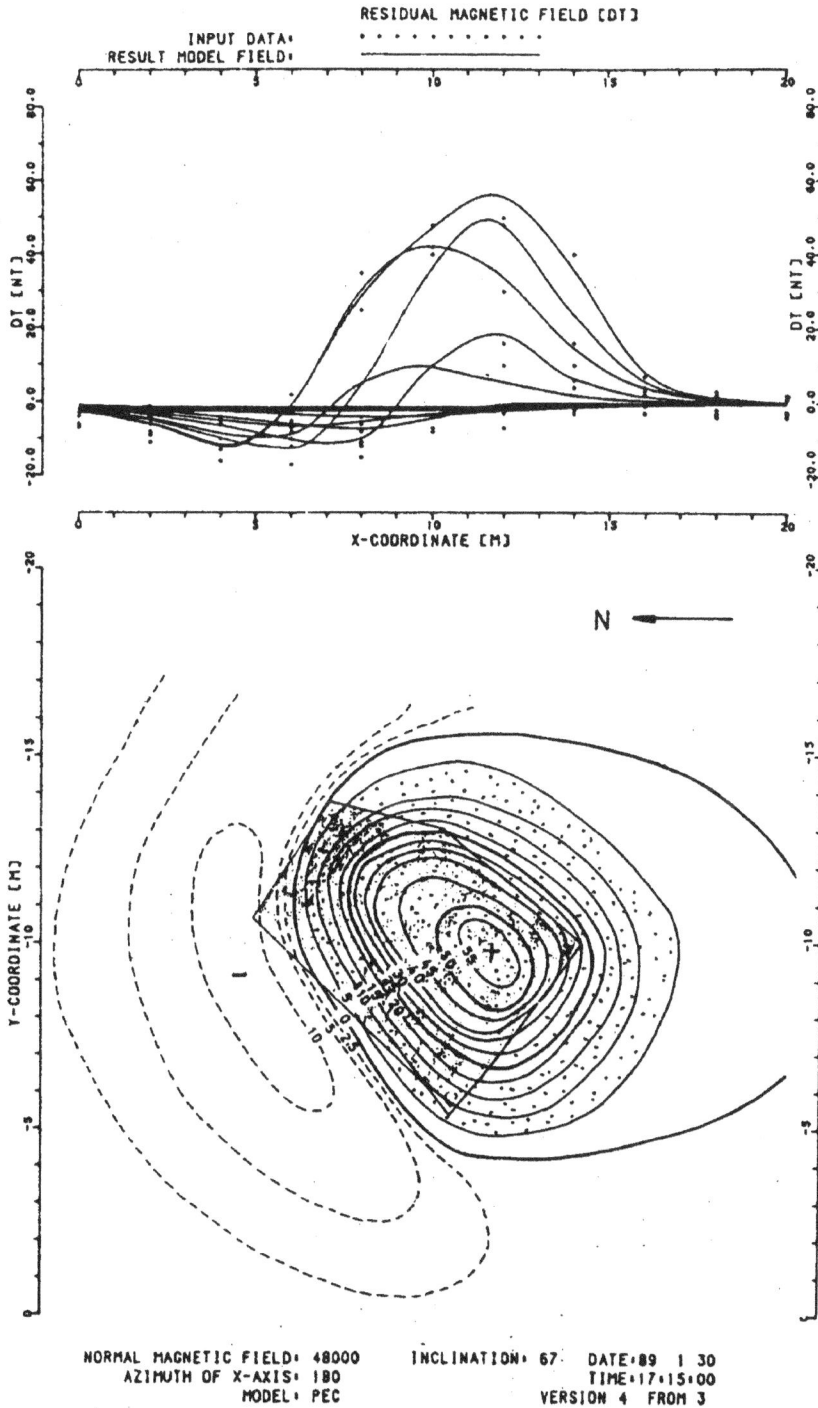

*Fig. 45. Areal processing of the resulting model in the space of the find of metallurgical furnaces dating back to the Late Roman time at Sudice, district Blansko (æ₁ = 47.9 . 10⁻³, æ₂ = 0.29 . 10⁻³ SI)*

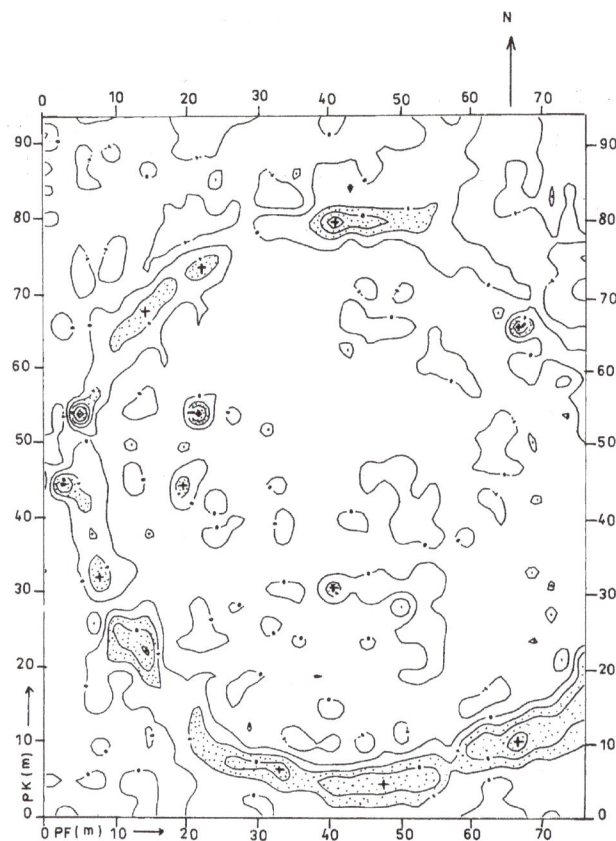

***Fig. 46.*** *A map of isolines $M^V(x)$ from the locality of Němčičky, district Znojmo*

Note: A new possibility for the quantitative interpretation of archaeogeophysical data arises by the application of the complete gradient method with the analytical continuation of the field to a lower level (Berezkin 1988). It is designed for the separation of inconspicuous ΔT anomalies and $\rho_a$ which are connected with the studied archaeological objects, and further for obtaining more detailed data about their sources. These anomalies are separated by means of filtering abilities of the operator $G_n$ which is a complex nonlinear filter. Its connection with the analytical continuation to a lower level allows to obtain a quantitative idea about the position of archaeological objects in the section as well as in the plane. The above information can further be successfully used in the solution of the reversed task, i.e. in finding the individual parameters of the studied 2 1/2D body. In the calculation of the complete standardized gradient with analytical continuation of the field the algorithm of the quick Fourier transformation was applied (Hašek - Vencálek 1991).

*Geoelectric Methods*

In geoelectric methods the quantitative interpretation is performed above all in the VES method. It is a determination of the thickness and the resistances of the individual layers of the section. To carry out the corresponding computer processing it is necessary to know the approximate estimate of the input parameters $(h_1....h_{n-1}, \rho_1...\rho_n)$. This can be done in several ways, mostly by interpretation by means of two- and three-layer theoretical curves. The program for the calculation of the direct task is based on the combination of the linear method of least squares for inappropriately

conditioned matrices with the Marquardt algorithm for solving the nonlinear problem of least squares. Such combination of parameters $(\rho_i, h_i)$ is looked for which best suits the measured data. In calculating $\rho_a$ curves of VES one starts from the Ghosh relation. The modified program used for the interpretation takes into account coefficients according to Nyman - Landisman (1977). The errors of the individual formulas are negligible in comparison with those of the measured values and inaccuracies due to the idealization of the model (horizontally layered environment).

McNeil (1980: 5-14) mentions the possibility of sounding and interpretation in the DEMP method under the condition of the existence of a two-layer environment for the case when between the two layers there is a great contrast in conductivity, so that $\sigma_1$ or $\sigma_2$ can be neglected with respect to the other layer. The method of graphical processing which is applicable under favourable conditions also for the purposes of archaeogeophysical prospection is based on the knowledge of $\sigma_2$ and $\sigma_1$ and in finding the thickness $z_1$ of the first layer, respectively. An inaccurate determination of $\sigma_2$ with the conductive first layer can even produce a considerable error in the final interpretation.

## 3.4. Physical Properties of Rocks Significant in Archaeological Prospection

For a successful employment of magnetometry and geoelectrical methods in solving various tasks of archaeological investigation and in the subsequent interpretation and/or the reinterpretation of the measured data physical properties of the objects of our interest and the surrounding rock environment are found despite the unfavourable situation when the archaeological objects are found prevailingly in Quaternary deposits (alluvium, loesses, blown sands, weathered material mantle, etc.).

Physical properties of rocks from the space of interest of the individual archaeological structures are determined by measuring of samples in the laboratory and in situ.

Rock samples taken from archaeological objects for laboratory measurements, can be studied in detail from the mineralogical point of view and texture. A certain disadvantage of this method of determining physical properties is the fact that the obtained results can be affected by the following factors: impairment of the natural deposition of the rock, change in its humidity, temperature, etc. Besides that, the sampling from areal excavations, dug and bored probes etc. is usually not carried out with such density (samples have little volume, they dilapidate) and representation (there is missing e.g. the effect of cracking of rocks, etc.) to clarify the quite evident character and the changes in all parameters along the whole section studied (Hašek - Měřínský1991: 21-32, 69-74). In geophysical measurements for the purposes of archaeological investigation, in the taken rock samples for magnetometry and geoelectrical methods (circumstances permitting), above all the following physical parameters are determined: magnetic susceptibility (æ) or the natural remanent magnetization (Mn) and the specific resistance (ρ). Changes in physical characteristics are connected mainly with changes in the material composition, inner structure and they are due to different geological factors, such as weathering, the degree of rock impairment, etc.

The measurement of physical properties directly in the field is carried out at a known archaeological structure or its exposed part, i.e. fortification, habitation object, gallery, etc. The obtained values often differ in these cases from the laboratory results. The dominant position in archaeogeophysical prospection is that of the field finding of physical properties by means of parametric measurements. The result is not necessarily a determination of the required parameter, it can also be some other kind of finding, such as a correlation between different parameters etc. (Mareš et al. 1983). A special task in the above works is that of the methods of continual measurement, such as in shallow pedological borings by means of which it is possible to obtain, after the elimination of negative phenomenona (diameter of the borehole, resistance of the outwash, etc.), the most complete overview of the character of the parameters studied under natural conditions. The results also include the effect due to the thicknesses of the individual near-surface layers, the arrangement of the measuring system, conditions of measurement, etc.

In the process of determining the physical properties of the rock environment from field measurements in shallow borings a complex of logging methods is applied - for our purposes we use mainly electrical and magnetic loggings. From the results of the processing, depending on the diameter of the borehole, data are obtained to the distance of about 30 - 50 cm from its wall which can also be applied to a wider surrounding of the individual boreholes. The method requires uncased boreholes. In case of small areal changes it is also possible to make a statistical evaluation of the individual parameters over the whole space of interest (the selective average, standard deviation, correlation dependences, etc.). With respect to the fact that the near-surface sectors of boreholes are usually cased, many times it is impossible to find the necessary data about the physical properties of cover and anthropogenic formations which are the main object of interest by logging methods, and for these reasons parametric measurements are applied. These works implemented at technical works (boreholes, probes, areal excavations, galleries, etc.) are used to find both electric and magnetic properties of the rock massif (rocks, flysh rocks, soils) and of archaeological structures (character, filling of the object, etc.).

From the evaluation of magnetic and electric physical properties of rocks determined from laboratory and field measurements can be drawn the following:

The size of magnetic susceptibility (æ) and of the natural remanent magnetic polarization (magnetization) ($M_n$) of the rocks depends on the content of ferromagnetic minerals (magnetite, titanium magnetite, titanic iron ore, pyrrhotine, etc.), on their chemical composition, dimensions, shape, percentual representation and on internal stresses in crystals and grains. In general the representation of magnetic

minerals in the rock increases with increasing basicity, i.e. in more basic bodies or archaeological structures composed of these rocks - diorite, gabbro, basalt, etc. higher values æ will be measured than in the acid ones (syenite, granite) which, however, need not always be the rule, cf. e.g. the results of the measurement of samples of Assuan granite in which æ = 5.57 . $10^{-3}$ SI (Hašek - Obr -Přichystal - Verner 1986: 149-187). For comparison it is possible to mention the found extent of æ in granites of the Třebíč, Jihlava and Dyje massifs which varies in the range of 0.06-0.2x$10^{-3}$ SI (Hašek - Měřínský1991: 28-29). The reduction of æ can occur also due to the positions of milonitized rocks, because in the course of crushing the rock there is a gradual change of magnetite to hematite - by martitization, which in turn also results in reducing the values of æ (Hašek et al. 1978).

The susceptibility of soils has a very close relation to the æ of the rock from which this component arises. Magnetite which is not subject to the processes of weathering, is in soils in a considerable percentual representation. Organic compounds in humus soils assist the rise of the magnetic mineral maghemite from nonmagnetic oxides of iron. The increase in æ and hence in the induced magnetic polarization ($M_i$) can be due to magnetic minerals contained in the near-surface layer; they also become the source of noise in detailed measurements since they form small aggregations in the microrelief of the measured area. The natural remanent magnetic polarization of the soil can be due to the heating of clays containing magnetite, but prevailingly due to viscous remanent magnetic polarization acquired gradually by the long-term operation of the magnetic field (Breiner 1973). This magnetization is current in the near-surface layers of soil and it is sometimes impaired by artificial intervention in this environment. The $M_n$ of soils can exceed $M_i$ as much as twice. From what has been said it follows that the soils can, in some cases, have even higher magnetic properties (e.g. alluvial silts) than their substrate or the archaeological object.

The most apparent anomaly of thermoremanent magnetic polarization (TRM) which is the most stable kind of remanent magnetization in minerals that have experienced heating (burned stones of furnaces, clays in the body of a defence line - vallum and from the destructions of overground objects, accumulation of stoneware, bricks, etc.). With this phenomenon there sometimes occur chemical changes in minerals. Ferromagnetic minerals can arise, or highly magnetic materials can change into less magnetic ones. Thus, at the temperature of 275° C the ferromagnetic maghemite changes into nonmagnetic hematite. The greatest TRM arises at the heating of pelites with a high content of magnetite, the smallest one in clays containing hematite (Frantov - Pinkevich 1973: 118-121).

In Table 2 there are extents of magnetic susceptibility of some archaeological objects occurring in comparison with the surrounding rock environment mostly by positive anomalies of the geomagnetic field (see Figs. 24 through 33). A number of complicating factors described in Chapter 3.5. can appear here.

**Table 2:** Magnetic susceptibility of some archaeological objects and building materials in the Czech Republic

| Object of Investigation | extent of magnetic susceptibility (u.$10^{-3}$ SI) |
|---|---|
| Open habitation | |
| - masonry of brick (unburned) | 0.5 - 2.1 |
| (burned) | 6.7 - 7.3 |
| Of sedimentary rocks | 0.0 - 0.4[1] |
| igneous rocks | 0.0 - 29[1] |
| Metamorphites | 0.0 - 44[1] |
| - burned houses with original loam and wooden and stone structure | 2.6 - 26.5 |
| - half-recessed and recessed huts, store pits, loam pits[2] | |
| - stake wells, circumference | 0.2 - 0.8 |
| Fortifications | |
| - loam and soil vallums, burned | 5.1 - 10.8 |
| - troughs, moats[2] | 0.6 - 1.3 |
| Burial grounds | |
| Grave pits | 0.1 - 1.3 |
| Production objects | |
| - fireplaces, sometimes filled with waste burned loam | 0.4 - 0.8 |
| - metallurgic furnaces + hearth debris, lining | 3.5 – 64 |
| - pottery kilns, bread ovens, etc. | 0.6 - 0.9 |

Notes:   1) the extent of æ can be even greater for the individual rocks (see Hašek - Měřínský, 1991: 31)
2) the filling of object can consist of naturally accumulated material from the surroundings together with the destruction of walls, secondary environment filled with ashes, daub, refuse from furnaces, pottery, etc.

The sizes of the specific electric resistance (further only resistance or resistivity) are sometimes affected by a number of natural geological and hydrological factors, such as the structure and texture of the rock, conductivities of rock-forming minerals, porosity, fracturing, fissures, mutual connection of cavities, their saturation with water, the degree of mineralization, intensity of rock weathering, etc.

The data about the sizes of resistance of the individual kinds of rock can be obtained mainly from the VES measurement (and/or from resistance profiling) near artificial excavations, boreholes, etc. Parameters determined from this processing and their employment for further works differ in some cases from the magnitudes found e.g. from electrologging measurement in shallow boreholes. Besides general anisotropy, i.e. the disagreement between the parameters of anisotropy according to VES and Ra are also affected by different arrangements of the measuring electrodes, such as the width of the depth intervention, etc. The found values can further be affected with areal inhomogeneity, weakened zones, etc. They correspond quantitatively better to the conditions of near-surface geoelectric measuring, even though data determined from the electrologging diagram yield more accurate data. It is evident that a greater agreement appears in petrographically better characterized types of rocks (Hašek - Měřínský 1991: 26-32).

The resistance properties of soils depend, according to Novotný (1973), mainly on their composition, moisture, salt content, concretion, temperature and to certain extent also on the ratio of sizes of grains they consist of. The author gives their general conductivity characteristic:

a) sandy soil dilapidates and does not form clods. Rain water thus quickly soaks it and the soil gets dry easily, it has reduced conductivity,
b) loamy soil maintains humidity, it is conductive,
c) clayey soil gets wet after heavy rains, it is therefore very conductive, if it gets dry, it is hard and little conductive,
d) calcareous clays (marls) are well conductive,
e) loam-sandy, clayey loam and sandy-loam soils are differently conductive according to the composition and moisture,
f) humus originating by the decay of organic remains in the soil is usually the most conductive.

The content of water (hydroscopic, capillary, seepage) in the near-surface layer depends on its composition, but particularly on the quantity of precipitation which can change the size of resistances. Thus, after a long-lasting rain the sign of the object can be suppressed by a low specific resistance (moat, recessed house) and a structure with high resistance (wall, stone wall, cavities, moat filled with stone destruction from the vallum body, etc.) can be located. The opposite takes place in a dry period. Similar properties also hold for measurements in different seasons of the year. With dropping temperature conductivity drops, and vice versa. Due to the complexity of changes in specific resistances in the near-surface layers and also in archaeological objects due to a number of external factors (processes depending of the time factor, solidification, etc.) a detailed analysis of the studied parameter is necessary for each individual case.

A comprehensive overview of physical properties of rocks from the whole region of Moravia and their relation to archaeogeophysical prospection is given in the paper Hašek - Měřínský(1991, 26-30). From those materials it follows that e.g. the Quaternary cover (alluvium, loesses) and the weathered material mantle (eluvium, deluvium) are, with respect to the lithofacial diversity, characterized on the whole by an extensive diversity of specific resistances and partly also
of changes in æ. From the viewpoint of physical parameters they can be divided roughly into two main parts. The first group includes all loamy, clayey and clay-sandy sediments characterized by reduced values of specific resistances and increased magnitudes æ (see Table 3). The second group includes sands, gravel-sands, gravels, talus (with different transitions and variable percentual representation of the loamy component) with increased values of resistances. Magnetic susceptibility for these cases is practically zero.

The general diagram of resistance differentiation of some archaeological objects is expressed in Table 4. Typical curves of apparent specific resistances found above different objects of interest are given in Figs. 33 through 43.

**Table 3:** Physical properties of the Quaternary cover and the weathered material mantle from data of field measurement

| Type of rock | Extent of physical properties | |
|---|---|---|
| | Specific resistance (ohmm) | Magnetic susceptibility (n . $10^{-3}$ SI) |
| Alluvium - alluvial loams | | |
|     - clayey loams | 12 – 60 | 0.33 - 0.90 |
|     - sandy loams | 80 – 120 | |
|     - sands, gravel sands | | |
|     - dry | 200 – 1500 | |
|     - wet | 40 – 250 | |
| Sandy loesses, loess, loessy loams | 10 – 50 | 0.10 - 0.15 |
| Eluvium – topsoil | 12 – 50 | 0.30 - 0.70 |
| Sand, rough sand, decayed substrate | 100 – 260 | 0.30 - 0.70 |
| Deluvium – slope loams sands, talus (different representation of loamy components) | 40 – 120  300 – 1800 | 0.20 - 0.40 |

**Table 4:** Resistance differentiation of some archaeological objects

| Archaeological Object | Surrounding Rock | Object Resistance in Relation to Surrounding Environment |
|---|---|---|
| cultural layer, habitation object with stones, pottery fragments, bones, pieces of charred wood | clayey soil sandy soil psamites, psefites, rock substrate | Increased Same Reduced |
| Fortification moat filled with soil moat filled with soil and stones | clayey soil, clay-sandy soil sandy soil, rock substrate | increased, reduced increased |
| Archaeological Object | Surrounding Rock | Object Resistance in Relation to Surrounding Environment |
| Grave (barrow) - filled with soil - wooden structures - stone structures | clayey soil clay-sandy soil sandy soil rock substrate | increased, reduced  increased, same |
| Foundation masonry Of stones | clayey soil, clay-sandy soil psamites, rock substrate | increased increased |
| Remains of mining activity - dry galleries - water filled galleries | clay-sandy soil rock (up to 10 ohmm) | Increased Increased Reduced |
| Shafts - sand and gravel filled - water filled - stone filled | clay-sandy soil rock(up to $10^3$ ohmm) clay-sandy soil | increased, same reduced increased |
| Other objects (cellars, loess, etc.) | rock (up to $10^3$ ohmm) clayey and clay-sandy soil rock substrate | increased, same increased |

Relative permittivity $\varepsilon_r$ is utilized in electromagnetic methods with higher frequencies (>100 KHz). The studied parameter expresses how much the condenser capacity increases if a certain rock is used instead of the vacuum. The practical $\varepsilon_r$ was tried at samples of coal (11-15), gneiss (8-12), clays and sandstones (3-6) (Hašek et al. 1983a; Záhora 1979).

## 3.5. Complicating Factors and their Effect on the Results of Geophysical Works

The relevance of using magnetometry and geoelectric methods and the credibility of results obtained in solving various archaeological tasks are conditioned by the low level of interfering effects of both natural and artificial origin (industrial interferences, erratic currents, etc.). Among interferences can be counted all circumstances that on the one hand reduce the possibilities of a successful application of geophysics in implementing the individual tasks, which brings along also negative effects on the productivity or measurement accuracy, and on the other hand they complicate the interpretation of the measured data (Hašek - Měřínský1991: 69-81).

The results of geophysical works can in essence be affected by three basic kinds of interferences of natural origin:
   a) the effect of the geologic structure (the so-called geological noises),
   b) the effect of archaeological situation,
   c) the effect of recent activity.

These factors must be considered and eliminated according to need in each task solved.

### 3.5.1. Effect of Geological Structure

The task of archaeogeophysical prospection is to find objects based prevailingly in Quaternary unsolified rocks. The only exception is perhaps the subsurface remains of mining activity, such as galleries, shafts, etc. Shallow geological formations can be, with respect to a great lithofacial diversity, considerably inhomogeneous both in the vertical and in the horizontal direction in comparison with its substrate consisting of Tertiary or pre-Tertiary rocks. These properties are negatively reflected in changes in physical parameters and thus also in the results of geophysical works. Thus, the size of specific resistances of the cover are, besides the lithological composition, affected by the variable granularity composition, porosity and the degree of saturation by ground water, its mineralization, etc. The resistance of the near-surface layer of loams (sandy, clayey) depends on their composition, on the presence of various granularity fraction and on the content of water. They usually have remarkably lower resistances than those of sands and gravel sands which are the highest of the above group, even though the clayey admixture makes them have only increased values of specific resistances. A similar situation also occurs during uneven water bearing. According to the values of resistances the aquiferous sectors can be interpreted as the "wedging" of layers of different lithological composition. Besides, a certain inhomogeneity of nonanthropogenic origin in the studied horizon (a rock lenticle of different physical properties etc.) can affect the whole interpreted magnitude of the specific resistance. Very problematic is also the separation of the archaeological object from the surrounding environment in case of its less different properties considering merely the results of the geophysical measurements (see Table 3).

As an example (Fig. 47) it is possible to present the results of the magnetic measurement over the habitation object at the Slavonic ringwall Pohansko (district Břeclav), where in the space of the SW suburbium the superincumbent bed of archaeological objects consist of the Dyje alluvium of increased thickness. Magnetic properties of the alluvia and the filling of the recessed object, as shown by later investigation, are approximately the same, which causes complications in the location of these objects under the above conditions. Only after removing the near-surface magnetic layer the habitation object exhibited a distinct anomaly of the geomagnetic field. (Hašek - Měřínský1991: 70, 123).

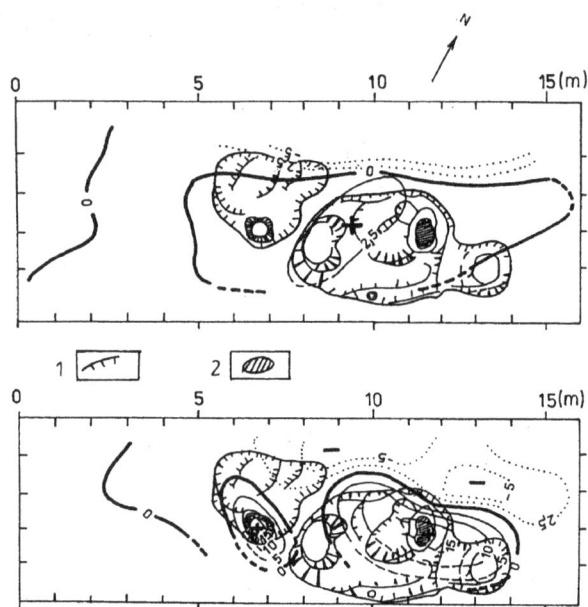

*Fig. 47. Břeclav-Pohansko, district Břeclav. Magnetic measurement above a Slavonic habitation object in the space of the southwestern suburbium 1- object delimitation, 2- furnace*

In geophysical practice one can often come across a case when above the followed body there occurs a thin layer with reduced specific resistance. By studying the effect of this near-surface conductive inhomogeneity in a homogeneous environment where a vertical nonconductive plate had been placed was dealt with by Dey et al. (1975: 566-572) and Nesterov et al. (1938). Their investigation proved that these "disturbing anomalies" have little effect on the three-electrode arrangement (AMN, MNB), whereas in the Schlumberger and the dipole arrangement is this effect considerable. From the results submitted by those authors it is possible to state that in the case of natural conditions of deposition (a thicker superincumbent bed) the near-surface cavities are difficult to locate by means of the standard electrode arrangement. For their more reliable detection it would be advisable to introduce newer variants of geoelectric resistance methods. Gupta - Bhattacharya (1963: 608-616) in

their calculations proved that the smaller the $\rho_1$, the higher is the share of the current of the focused arrangement in the lower semispace $\rho_2$, from which there also follows the suitability of the application of differential arrangements of electrodes (MAN, AMNA') for a more detailed precisioning of the course of cavities and other nonconductive bodies of anthropogenic origin below the lateral inhomogeneity near the surface.

Some mechanical and lithological properties of the rock massif studied for the purposes of archaeological investigation are not always reflected in the results of geophysical works. That applies e.g. in tracking the interface between the weathered and the compact rock or the relief of the crystallinicum and its superincumbent beds, loess horizons, etc. It can be due to both the intensity of geological and hydrological factors (the degree of mineral weathering, fracturing, water content etc.) and the peculiarities of the spatial orientation of the studied layer (depth, thickness, etc.). From that can be deduced the objective limitation of the applicability of geophysical methods for the required objectives of prospection.

In many cases there is another extreme, when the object of investigation particularly differs from the surrounding environment in its physical parameters (such as magnetic properties, specific resistances, etc.) and the measured values separate anomalous bodies that could be of archaeological origin, but the investigation is quite negative in this point.

As an instructional example it is possible to submit the results of geophysical measurement and the subsequent archaeological verification at the locality of Diváky (district Břeclav).

By observing from an elevated place in the field, in the area east of Diváky (district Břeclav), on a mild southern slope of a field opened to S a hint of a circular structure was located. From the results of magnetometric measurement at an area of 130 x 100 m in a grid of 2 x 2 m in the form of a map of isanomales ΔT (Fig. 48) two intense anomalies were separated in that space (+20 nT) with the axis approximately in the NE-SW and NW-SE directions, accompanied by a number of local isometric anomalies, particularly in the neighbourhood of the larger structure of about 20 x 10 m, located in the centre of the area of interest. (Bálek - Hašek - Měřínský - Segeth 1986: 568-569; Hašek - Měřínský1991: 71-74).

On the basis of complementing geoelectric resistance measurement (SRP, VES) and sounding with the pedological pole (Zelenka 1985) in the area of the NE ΔT anomaly (Fig. 49) it was found that its source was a dark filling of increased æ and specific resistances, reaching the thickness of as much as 2.5 m. The areal extent of this anomalous body is about 50 x 15 m (Hašek - Měřínský1991: 73-74).

Further sounding works proved that the source of magnetic anomalies and apparent specific resistances was a morphological depression in the sector of yellow-brown sandy loesses filled with homogeneous humuos displaced blacksoil, i.e. that it was a geological matter - an aspect of soil changes and not a phenomenon due to man's activity (Bálek-Havlíček 1987).

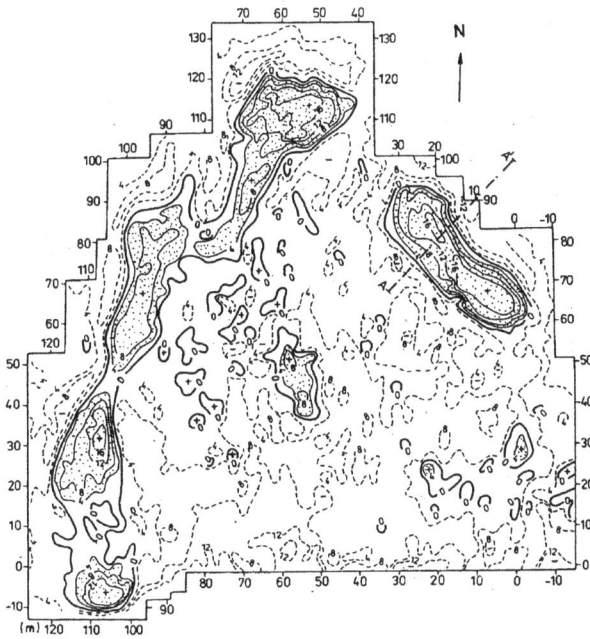

*Fig. 48. Diváky, district Břeclav. A map of isoanomales ΔT in the studied space and the situation of the verifying profile A-A'*

Similarly, an ascent of magnetically anomalous rocks covered by a thin layer of weathered material can sometimes be erroneously interpreted as an archaeological object.

### 3.5.2. Effect of Archaeological Situation

The methodology of geophysical works and the interpretation of the measured data are affected by both physical parameters of the studied objects and their size, depth and location in the rock environment. A considerable effect on mapping the object in the field (a moat, a habitation object, a grave, etc.) is that of its physical distinction from the environment (see Table 3).

In the case when e.g. a depression structure due to man's activity is filled with the same dug out material, zero anomalies of the magnetic field are measured, but the values of the apparent specific resistance are different in comparison with the surrounding environment. The filling consisting of blown or outwash material from the surrounding surface layers together with wall destructions of the object walls exhibits smaller ΔT anomalies and increased or decreased values of $\rho_a$. The humus filling with remains of organic materials and Fe-objects is reflected by a more intense ΔT anomaly, for $\rho_a$ apply similar properties as in the preceding case. Analogous properties exist in finding stone or brick masonry. In the foundation masonry located in grave sand and covered with stone destruction from the overground parts of the object of greater thicknesses the possibility of its

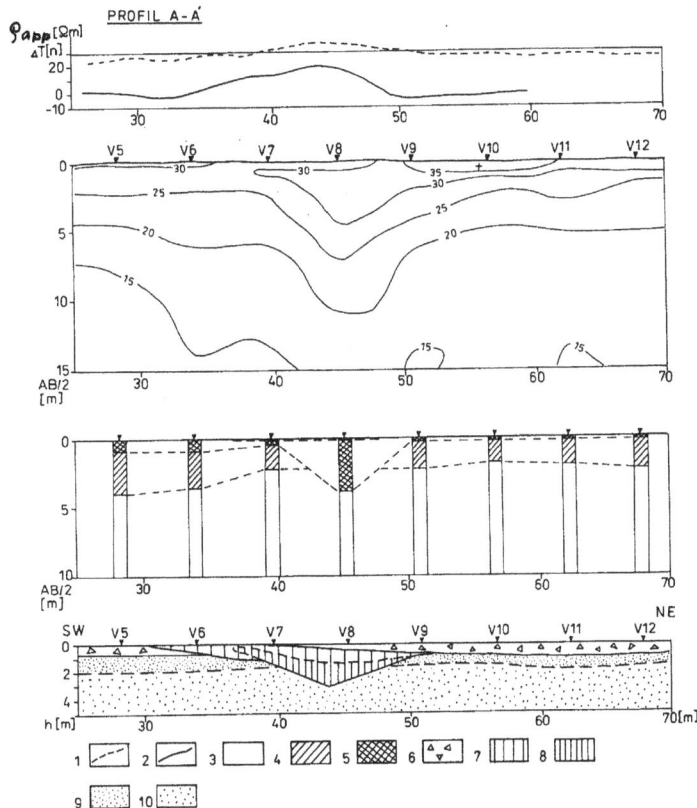

*Fig. 49. Diváky, district Břeclav. Results of geophysical measurement and pedological verification at the A-A' profile, according to Zelenka (1985) 1- SRP curve at the geometry A2M2N2B, 2- relative values of ΔT, 3- resistance of layer 20 ohmm, 4- resistance of layer 20 - 30 ohmm, 5- resistance of layer 30 ohmm, 6- topsoil, 7- filling of depression (lighter czernozem), 8- filling of depression (darker blacksoil), 9- loess with loamy percolations, 10- loess*

location (e.g. by the method of resistance profiling), is mostly considerably limited. Also the tracking of the foundation ditches left after removed masonry is under similar geological conditions more complicated (Hašek - Měřínský 1991: 74-75).

From the above it can be deduced that by geophysical methods it is possible to directly find archaeological objects characterized on the one hand by differently coloured filling than that of its surrounding, i.e. moats, habitation and settlement recessed objects, graves and further structures characterized by increased æ and different values (increased, reduced) of specific resistance from the surrounding rocks, further stone structures (foundation masonry, elements of architecture) with increased specific resistances, objects of burned material (fireplaces, furnaces, daub, burned clays from the body of a vallum, etc.) and, finally, also bodies with higher content of iron - metallurgical installations (æ>>0). Indirectly is it possible to locate e.g. graves filled with the same material in which a large Fe-object had been placed (a sword, a lance, etc.), habitation objects with a filling of the same colour as the surrounding of the located furnace, etc.

A further effect on the results of geophysical works is that of the object size and the depth of its foundation. From this point of view the finding of e.g. stake holes from overground objects is sometimes quite issueatic for geophysical works, despite the fact that æ>0, (the size of the anomaly of the magnetic field drops in a three-dimensional body with the cube of the distance). For their possible location it would be necessary to carry out very detailed measurements with the network of the order 0.5 x 0.5 m, or even more detailed (see Chapter 3.1.).

The deformations of the anomalous image in the Wenner profiling due to different nonconductive bodies were dealt with by Kyono (1950: 29-59). He calculated the profile curve of the above configuration from the theory of conform images. He points to different ways of obtaining the results for the case of a hollow cylinder. By analysis he found that it was possible to find a cylindrical cavity under optimum conditions using a current electrode geometry up to the ratio 5 : 1 (medium depth/ratio). It is, however, necessary to bear in mind the assumption that the above ratio applies only for ideal geologic and geometric conditions and dimensions of the given interfering body. In Table 5, compiled according to Lösch-Militzer-Rössler (1979: 53-126) the location of a nonconductive cylindrical body is given for an objective approach (a cavity, masonry, etc.) at different depths by means of the common electrode arrangements (the significant limit = 0.1 A.E.).

Geophysical methods applied in e.g. the solution of the issues of cavities (galleries, shafts, cellars, unvaulted cellars in loess, etc.) also represent an indirect method of their location. For a convenient explanation of measured data some preliminary information about the position, geometry and the depth of the body is necessary. That applies chiefly for microgravimetry or magnetometry in which mathematical modelling and calculation of corrections are essential. Due to diversity of structures (historical built-up area, remains after mining, etc.) and the appearance of cavities the necessary complementing data are not known in most cases, which can

sometimes cause considerable location problems (Hašek et al. 1981).

**Table 5.:** The Anomalous Effect of a Nonconductive Cylindrical Body in a Homogeneous Semispace With a Common Electrode Arrangement

| Electrode arrangement | Depth of the anomalous body centre(m) | | | |
|---|---|---|---|---|
| | h = 2 | h = 3 | h = 4 | h = 5 |
| Wenner – AMNB | + | + | | |
| Schlumberger – AMNB | + | + | +? | |
| three-electrode | | | | |
| - potential | + | + | +? | |
| - gradient | + | + | + | |
| dipole axial | + | + | + | +? |
| central gradient | + | + | + | +? |
| Differential | | | | |
| - potential | + | + | + | + |
| - gradient | + | + | + | + |

Mutual effects of a number of objects of several cultures at the so-called polycultural localities where the settlement existed for a long period of time from prehistoric to early medieval development can be included among further complicating factors of geophysical works. It includes objects of different cultures found side by side or they are in mutual superposition. Gradually a complicated stratigraphy was growing there. It is above all the location of overlapping objects from different periods of time, of various character and size. From that viewpoint an older moat impaired by a younger habitation object is characterized by a more complicated course of the magnetic field which need not be correctly interpreted as an effect of two independent objects. A similar situation can occur in recessing a younger grave into an older habitation object, etc. In many cases an Fe-object located either in the near-surface layer (recent affair) or in the studied object can fully overshadow the effect of that structure which in turn does not necessarily appear in the results of magnetometric measurement. Certain complications and the ambiguity in interpretation can occur e.g. in the case of applying geoelectric methods in solving the task of locating cavities, foundation masonry, etc., if complementing data about the character of the studied bodies are not available (from the areal excavation, boring works, etc.). Locally increased values of specific resistances can be due to both a built-up area, a near-surface cavity, and another inhomogeneity in the rock massif (silification, reduction of rock impairment, etc., cf. Hašek - Měřínský1991: 79-80).

### 3.5.3. Effect of Recent Activity

The sphere of factors can include mainly

    a) later interventions into archaeological objects to which various excavations are counted, deep tillage, etc., disturb the overall character of the measured field, which results in the fact that above the quoted environment (a filled pit, a drainage system, etc.) further anomalies are measured of both the magnetic and the electric field,

    b) the course of pipelines, cables etc. considerably complicate the employment of all applicable methods in the above prospection. Thus, in magnetometry the effect of those bodies can be so great that it suppresses completely the possible expression of an archaeological structure near them. An analogy also applies for the use of geoelectric methods. Above those bodies mostly a marked decrease in the specific resistances is measured (conductor) and the possible archaeological object is completely shadowed by the above type of inhomogeneities.

    c) recent Fe-objects or their greater concentration in the near-surface layer are disturbing effects for the positive application of magnetometry in solving the individual archaeological tasks. They represent a source of relatively intense anomalies and in places they also cover the effect of the investigated objects.

Man (Bárta-Man-Mašková 1985), in connection with the detection of unexploded ammunition, dealt with the problems of finding isometric Fe-bodies. He solved the anomalous effects of soft magnetic bodies on theoretical level, above all ellipsoids of high permeability. The resulting anomaly grows linearly with the mass of the body and its maximum value is inversely proportional to the cube of the depth. Further, he also expressed anomalies for bodies limited externally and internally with ellipsoidic surfaces, the so-called "shells". The calculations show that anomalous effects depend mainly on the external dimensions of the "shell", its orientation on the Earth's magnetic field and the place of observation. The thickness of the "shell" (anomalies $\Delta T$ are almost identical with anomalies for full ellipsoids) and the exact value of its permeability are only marginally important. For an ellipsoidic shell of external dimensions 1 x 0.2 m it was stated that the maximum differences of anomalies corresponding to the change in the depth of deposition of the centre of that body from 2.5 m to 3.5 m $\Delta T = 74$ nT for the case of the vertical orientation of the main axis of the shell, $\Delta T = 20$ nT for the horizontal orientation of the main axis in parallel with the magnetic meridian, $\Delta T = 10$ nT for horizontal orientation of the main axis perpendicular to the magnetic meridian.

## 4. RESULTS OF COMPREHENSIVE INVESTIGATION AND THEIR DISCUSSION

The following paragraphs of this chapter deal with practical experience and results achieved by archaeogeophysical prospection in a number of localities from the regions of the Czech Republic, Germany and Egypt in solving the individual tasks of archaeological investigation.

Most attention was paid to

a) the optimization of the methodology of geophysical works (in magnetometry and geoelectric methods) in searching for objects of different character (habitation and production objects, fortifications, etc.) and the time,

b) the development of the corresponding physico-archaeological model of the structure of interest (open settlement, fortified locality, burial ground, production centre, etc.) for the subsequent investigation in the field.

c) the comprehensive geophysical and geological processing and the comparison of the obtained results with the situation found by excavation works. For clarity we state the situation of the individual processed localities in the geological map of the Quaternary cover and the weathered material mantle of the CR (Fig 50).

*Fig. 50.* *1- Běhařovice, district of Znojmo, 2- Bohuslavice by Kyjov, district of Břeclav, 7- Břeclav-Pohansko, 8- Býčí skála by Adamov, district of Blansko, 9- Bylany, district of Kutná Hora, 10- Dalskbát, district of Olomouc, 11- Doubravník, district of Žďár nad Sázavou, 12- Drvalovice, district of Blansko, 13- Hodonín-Mikulčice, 14- Hory, district of Třebíč, 15- Jaroměř-Josefov, district of Náchod, 16- Jedovnice, district of Blansko, 17- Jemnice, district of Znojmo, 18- Jihlava, 19- Kokory, district of Blansko, 21- Kurdějov, district of Břeclav, 22- Lelekovice, district of Brno-venkov, 23- Lochenice, district of Hradec Králové, 24- Mikulov, district of Břeclav, 25- Mušov, district of Břeclav, 26- Němčičky, district of Znojmo, 27- Nový Jičín, 18- Olomouc, 29- Olomučany, district of Blansko, 30- Pavlov, district of Břeclav, 31- Podolí, district of Brno-venkov, 32- Předklášteří-Tišnov II, district of Brno-venkov, 33- Rašovice, district of Vyškov, 34-Rokštejn, district of Jihlava, 35- Sudice by Boskovic, district of Blansko, 36- Šitbořice, district of Břeclav, 37- Šumice, district of Znojmo, 38- Šumice, district of Uherské Hradiště, 39- Tábor, 40- Troskotovice, district of Znojmo, 41—Uherské Hradiště, 42- Valtice, district of Břeclav, 43- Vážany nad Litavou, district of Vyškov, 44- Vedrovice, district of Znojmo, 45- Velešovice, district of Vyškov, 46- Velké Bílovice, district of Břeclav, 47- Velký Újezd by Moravské Budějovice, district of Třebíč, 48- Vlasatice, district of Znojmo, 49- Vratíkov by Boskovic, district of Blansko, 50- Vyškov, 51- Znojmo*

## 4.1. Investigation of Open Settlements

The first sphere of solving these issues includes settlement and habitation objects without remains of foundation masonry connected with binders (mortar etc.), i.e. simple recessed buildings of different sizes, such as semi recessed and recessed huts, storage and refuse pits, loam pits, overground objects with pole structure, furnaces, cultural layers, etc. Objects of this kind are from late Paleolithic period characterised by fillings containing ashes, charred pieces of wood, stone artifacts and animal bones, from the Neolithic period by pottery, daub, slag, tools, stones etc., which differ by their physical properties (particularly magnetic) from the surrounding medium (see Table 2).

The second part of these works includes the investigation of remains of settlement, farming and other objects of different ground plans with remains of masonry (stone, brick burnt and dried), bound with mortar, and/or searching for burned layers of daub, signs of destructions from original buildings, etc.

### 4.1.1. Late Paleolithic Stations

*Dolní Věstonice - Pavlov, district Břeclav*

Rescue and advanced investigations of the AI CSAS Brno (now AI AS CR Brno) have shown that the overall structure of this Late Paleolithic station (Pavlovien) situated on the NE slope of the Pavlovské Hills, subjected to gradual thatch-shaped landslides and block slides is more complex than originally expected. According to hitherto investigations the divisional planes of the slides are situated into places of contact between Pleistocene sediments and the Pouzdřany clays forming an impermeable layer to seepage water (Klíma 1986). In the central part of the station an evident edge of the block slide passes in a slanting way along the slope. Its course within the reach of the upper part of the locality is not sufficiently apparent (Hašek - Měřínský 1991: 89-91).

The task of geophysical works at six perpendicular profiles was to delimit the lines of the planes of division and the thickness of Pleistocene sediments (sandy loesses, intercalated beds of blown sands, at the base limestone talus) with a cultural layer (Hašek - Ludikovský 1977a: 111). For solving the above task geoelectric methods were used (SRP, VES) and for completing the data also the method of shallow refraction seismicity.

The result was the delimitation of three essential zones of reduced specific resistances which can be related to the planes of division of the individual slides. The compiled geological-geophysical profile of the territory of interest according to geological data and the VES interpretation is represented in Fig. 51.

The magnitude plan of Pleistocene sediments, drawn according to the VES materials (Fig. 52) can be compared with the magnitude of the individual landslides and their planes of division. The interpreted depth without Pleistocene varies between 2 - 11 m in the processed region, which is in accordance with the sounding works carried out.

From the above data it can stated that the cultural layer spreads in relatively different depths, being impaired by a number of planes of division. Different age of the individual landslides can be excluded.

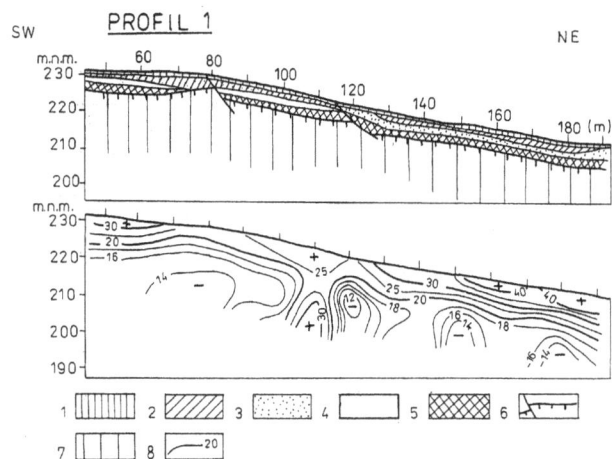

**Fig. 51.** *Pavlov, district Břeclav. A geological-geophysical section from the results of processing by the VES method 1- topsoil (50 - 60 ohmm), 2- loesses with major positions of blown sands (20 - 40 ohmm), 3- loess with high humidity (12 - 25 ohmm), 4- dry, sandy loess (40 - 50 ohmm), 5- solifluctional horizon with cultural layer and areas of gravels (38 - 80 ohmm), 6- root position, 7- clays of the Pouzdřany layers (NE Eocene-egeran) (11 - 26 ohmm), 8- isolines $\rho_a$ of the VES method*

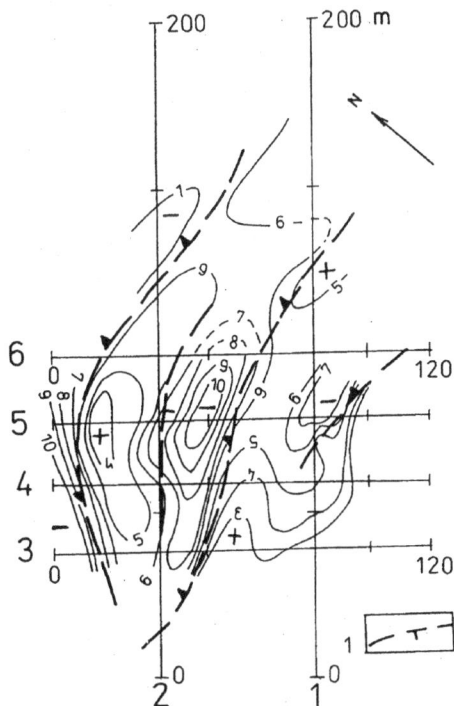

**Fig. 52.** *Pavlov, district Břeclav. A map of thicknesses of Pleistocene sediments 1- interpreted disturbances*

## 4.1.2. Unfortified Settlement Formations from the Prehistoric to the Slavonic Period

*Bořitov, district Blansko - "Na kříbě"*

In the cadaster of the community of Bořitov six settlements of the Bell-Shaped Goblet Culture people were noticed by surface collections. At three of them small probe investigations have been carried out. The field "Na kříbě" was measured magnetometrically in 1976 (Hašek-Ludikovský et al. 1977; Hašek- Ludikovský, 1977a) and in 1977 (Hašek et al. 1977: 14). The purpose of the measurement was to find the position, dimensions and contours of the assumed settlement objects.

In 1976 an area of 10 x 15 m was experimentally measured in a network of 1 x 1 m, in 1977 that of 50 x 80 m at the same step of measurement.

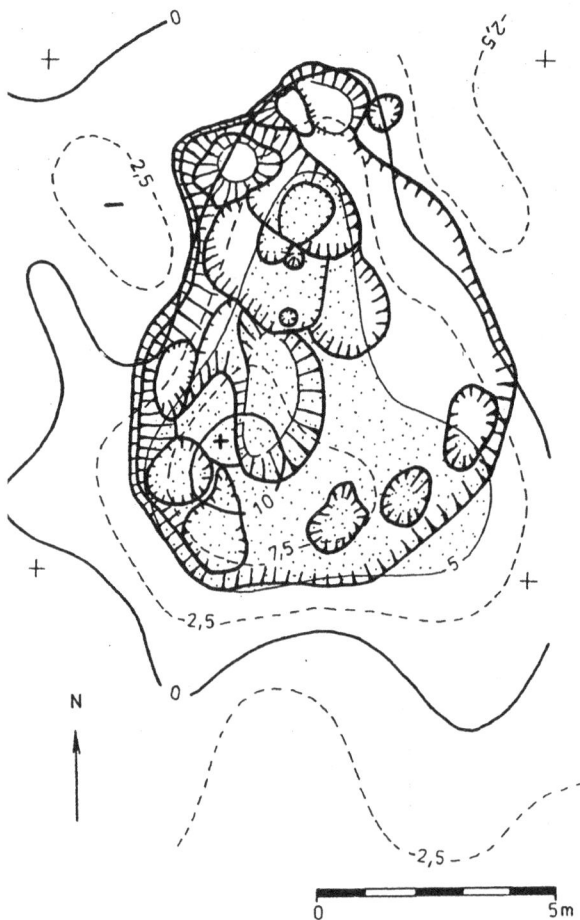

**Fig. 53.** *Bořitov, district Blansko. "Na kříbě". Map of isanomales ΔT and an object exposed in 1976.*

The results in the form of plans of isanomales ΔT and the situation of the exposed objects are submitted in Figs. 53 and 54.

From Fig. 53 it is evident that the source of the magnetic field anomaly (+ 12 nT) is a settlement object of the dimensions 7 x 10 x 0.9 m, recessed into the loess bed which, as appeared on the basis of the subsequent investigation by the AI CSAS in Brno, belonged to the Bell-Shaped Goblet Culture (Ondráček 1978).

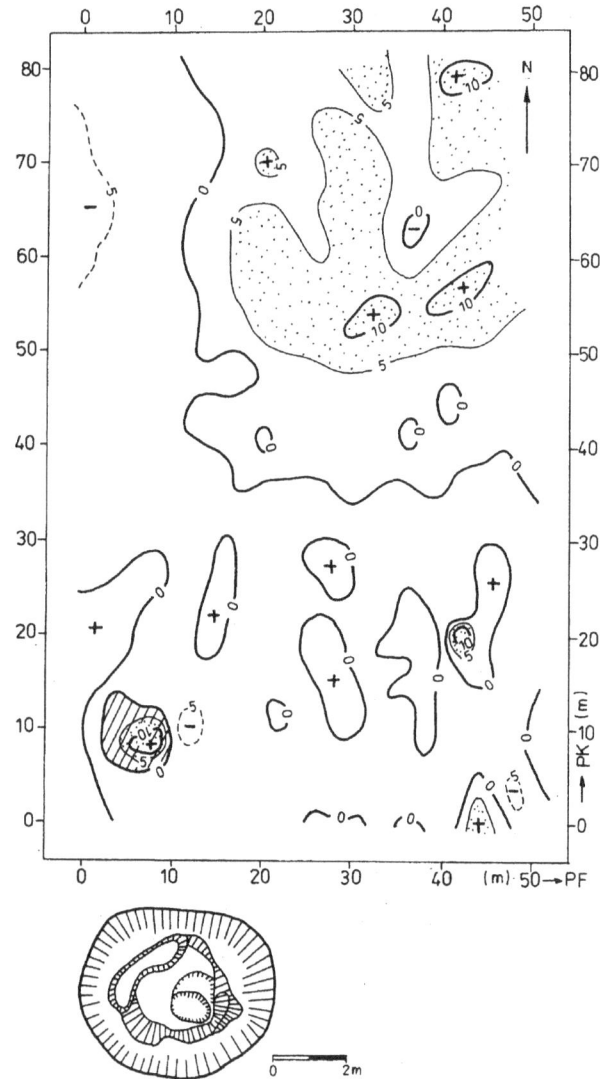

**Fig. 54.** *Bořitov, district Blansko. "Na kříbě". Map of ΔT isanomales and an object exposed in 1977.*

In the area measured in 1977, situated N of the 1976 excavation (Fig. 54), three intensive anomalies - marked A, B, C can be separated from the plan of isanomales ΔT, of which only in the case of isometric anomaly ΔT (+ 12 nT, marked C) it was possible to judge at the position of a settlement object of elliptic shape of the size of about 4 x 5 m. By archaeological investigation in this area an oval dish-shaped settlement feature was found of the dimensions of 5.2 x 4.4 m, with recesses in the lowest parts of the bed (see Fig. 54). The filling is dark brown to black, containing pottery and further archaeological artifacts, allowing it to be included into the Bell-Shaped Goblet Culture. (The source of the anomaly of A was found to be a recent Fe object, the anomaly of C is formed by a lithological change in the near-surface layer) (cf. Hašek-Měřínský 1991: 91-92).

On this example it is also possible to demonstrate the economic advantage of the application of magnetometry in solving the above task of archaeological prospection, because the distance of the two objects is approximately 37 m. Their ground plans found on the basis of the results of geophysical works made it possible to trace out the extents of the excavations and the course of the comparison profiles. On the basis of the interpreted depths of objects the volume of the field probes was estimated and hence the extent and the costs of the research.

*Vyškov, district Vyškov*

The rescue archaeological investigation situated into the area of the motorway intersection at the western boundary of Vyškov in the Carpathian Neogene foredeep was oriented on the settlement of the Roman Period (the 2nd - 3rd century A.D.) at the site of a mild slope inclined towards the SE with the height of 260 m above the sea level, along the left bank of the Rostěnický Brook (Fig. 55). Besides the Roman Period settlement was also discovered a late Slavonic settlement.

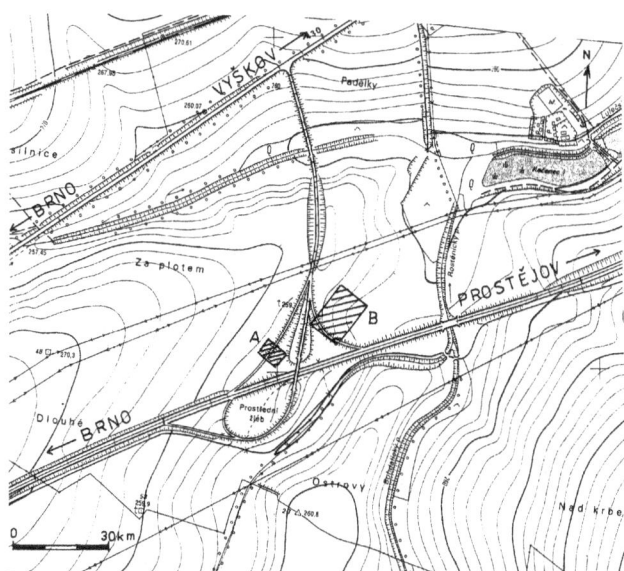

**Fig. 55.** *Vyškov. Scheme of the searched area.*

On the basis of areal excavations carried out there in 1988-90 by the former AI CSAS in Brno fundamental information was obtained about the extent and character of the settlement, particularly from the Roman Period, above all in the southwestern and the southern parts of the settlement (Šedo 1991). The investigation located and studied 105 objects of the Roman Period (recessed huts, stakehole buildings, storage pits, production and farming buildings, loam pits, furnaces, etc.) and further 20 buildings of a hitherto undetermined age.

Magnetometric measurements carried out in 1990 (Hašek et al. 1990) NE of the hitherto investigated area was to determine the positions of archaeological objects at two areas of the dimensions of about 65 x 50 m (detail A) and 140 x 100 m (detail B) in a grid of 2 x 2 m and thus provide more detailed information about the concentration of the settlement in the studied area. From the plan of isanomales $\Delta T$ in Sector A (Fig. 56) it was found that all major exposed objects of the dimensions of about 1 x 1 m to 6 x 6 m appear in the results of geophysical works. In the central part of the processed territory several more local isometric $\Delta T$ anomalies were specified in which no archaeological objects had been found by the investigation. It might be caused by

the fact that their source is either a lithological change or a thicker position of the cultural layer in the overlying beds of loesses scraped by a bulldozer in the excavation works. In all other cases a good correlation between the objects of the investigation and the anomalies of the magnetic field is evident.

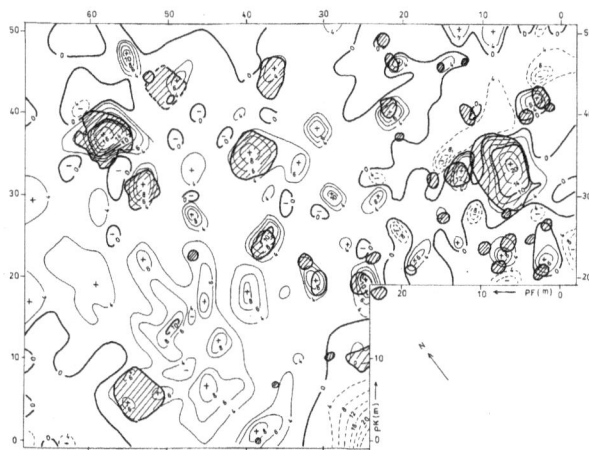

**Fig. 56.** *Vyškov. Map of $\Delta T$ isanomales (detail A) and the situation of the excavated objects.*

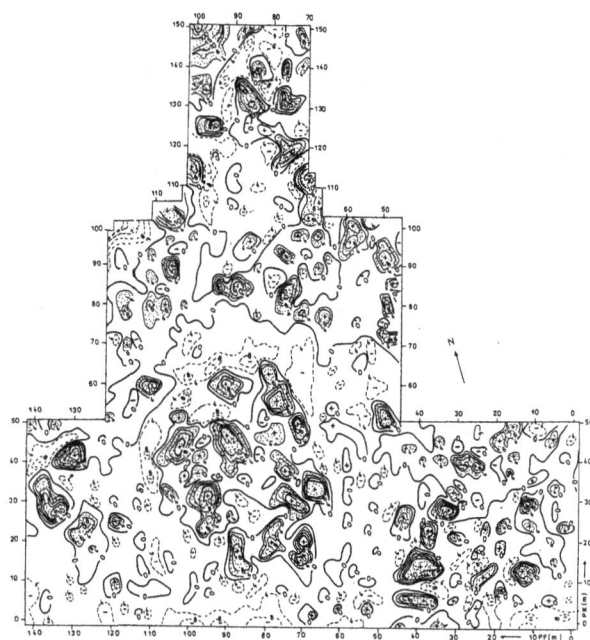

**Fig. 57.** *Vyškov. Map of $\Delta T$ isanomales - detail B*

In the case of sector B (Fig. 57), according to the distribution of $\Delta T$ anomalies (up to + 30 nT) a high regular concentration of settlement was found, in places with an apparent arrangement of objects into a circular formation. From the character of $\Delta T$ curves obtained in the measured area it is possible to pinpoint more than 70 different recessed objects (oval, square, rectangular) of settlement and other objects of various dimensions (about 2 x 2 m to 7 x 7 m) and ground plans. The character of the filling (cultural layer, slag, etc.) and their different depth of recession make the magnetic anomalies have a variable amplitude.

On the basis of the present state of knowledge it cannot be assumed that between the investigated part and the archaeologically uninvestigated part of the station there appears a significant drop in the density of the built-up area, i.e. that to the NE from the exposed area there is a region with a lower intensity of settlement.

*Uherské Hradiště - Staré Město "Za zahradou"*

In 1977-78 the Museum of Moravské Slovácko at Uherské Hradiště in cooperation with the Moravian Museum in Brno carried out a rescue investigation of the Slavonic and medieval settlement in the area of private gardens between the streets Školní and Jezuitská and the Hrdinů Square at Uherské Hradiště - Staré Město. This was due to a new building activity (Snášil1979, 1980; Snášil-Procházka 1980). The geophysical measurements were carried out by means of magnetometry at two selected areas of the dimensions of about 30 x 40 m (Hašek-Ludikovský et al. 1977: 20-30) and 40 x 20 m in the grid of 1 x 1 m (Hašek-Ludikovský-Mayer-Ondráček-Pantl 1979: 19-21). The purpose of the works was to find the density, mean size and distribution of the individual objects recessed into clayey loesses, thus contributing to the determination of the procedure or speeding up of the archaeological research.

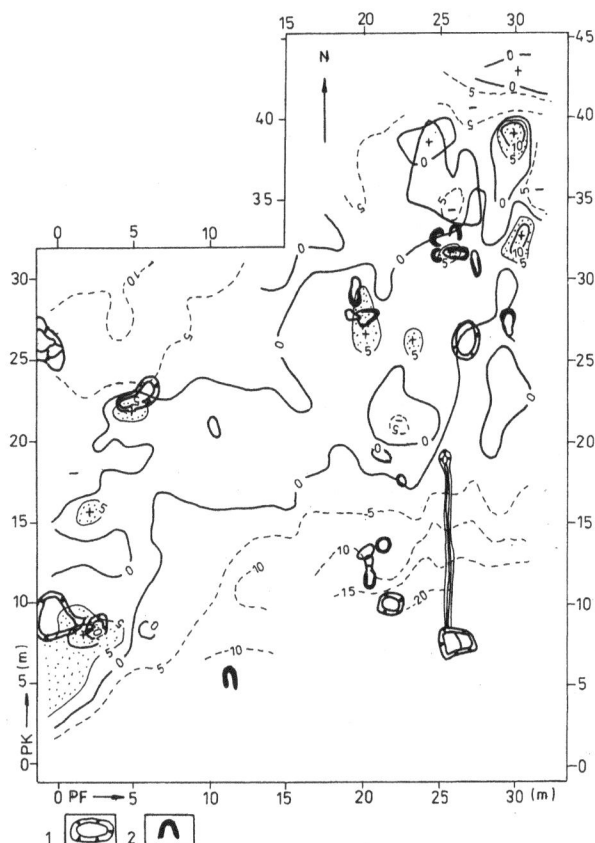

*Fig. 58. Uherské Hradiště - Staré Město "Za zahradou". Map of ΔT isanomales and objects excavated in 1977.*

From the drawn plans of ΔT isolines (Figs. 58 and 59) and the subsequent excavations it was deduced that the overall course of the magnetic field in the area of interest was impaired by the occurrence of objects of archaeological origin (recessed buildings, loam furnaces) and recent Fe-objects in the near-surface topsoil layer.

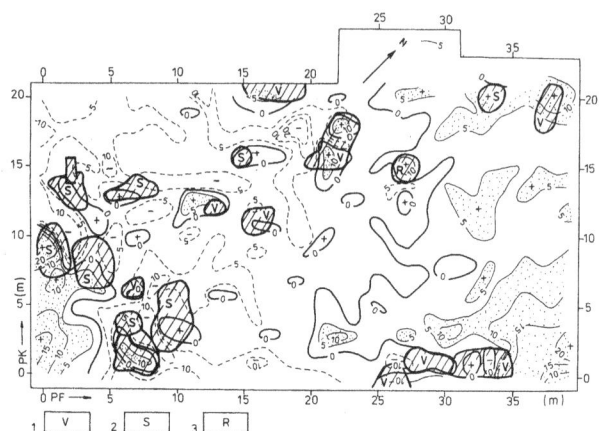

*Fig. 59. Uherské Hradiště - Staré Město "Za zahradou". Map of ΔT isanomales and objects excavated in 1978. 1- objects from the period of Great Moravia, 2- objects from the Medieval Ages, 3- recent objects*

From the area measured in 1977 (the SE part of the station) the total of 20 archaeological recessed objects was found by archaeological works (Fig. 58), 4 of the objects belonging to the Great Moravian period, 15 dating back to the 13th to 17th centuries and one of them being recent. Five objects, i.e. 25 %, were not found within the area studied by geophysical measuring. To the Slavonic period belonged e.g. a grain pit which was situated in the central part of the measured area, the diameter of its neck being less than 1 m, (see Fig. 58). From the medieval objects included in the period starting with the 13th century the remaining part of a medieval loam furnace was not found (object No. 8) and a narrow urine groove of a shed with sewage pits (object No. 11), at the S and W margin of the area studied remains of five loam furnaces (objects Nos. 15-16). On the other hand, in the reverse revision of anomalies obtained in geophysical prospection a shed and a cote with a sewage pit found (objects Nos. 17-18, cf. Snášil1980a: 136). The measurement in that sector was considerably difficult due to the existence of wire mesh which affected the measurement of the archaeological object.

By archaeological investigation of the area measured in 1978 (NE part of the station) (Fig. 59) the total of 18 objects was exposed, out of which 10 of the Great Moravian period, 7 of the period of the 13th to the early 17th centuries, and one recent object. Out of that number geophysically less distinct is only the demonstration of medieval objects No. 139 and 140, i.e. 88.8 % of objects were found by magnetometry. In the revision based on the plan of anomalies compiled on the basis of geophysical measurement, on the other hand one of the six kilns was found and subsequently exposed (object No. 144) belonging to the lime manufacture plant of the 13th century (Snášil1980, 1980a: 136). The sources of further ΔT anomalies are prevailingly local inhomogeneities in the near-surface layer (recent Fe-objects, fillings after extracting tree stumps, etc.).

Besides the overall density and situation of archaeological objects the geophysical works carried out at the above station determined also the direction of the spread of settlement concentration with the axis NE - SW which continues in this direction even outside the area processed.

From the submitted example of the application of geophysical methods in complex field conditions of the present-day urban built-up area it was deduced that it is possible to use magnetometry successfully even in solving the above type of tasks (cf. Hašek-Měřínský 1991: 94-96).

### 4.1.3. Extinct Medieval Villages

An important source of knowledge of the life in the Middle Ages is the archaeological investigation of extinct villages. It is the study of the life of peasants who formed a very numerous part of the population at that time. The investigation of extinct medieval villages brings material sources for the study of socio-economic relations, an instruction about the life conditions, environment, the development of material production and the types of village habitations (Hašek-Kovárník 1996). In the territory of the Czech Republic there are suitable conditions for comparison thanks to systematic investigations of extinct villages, such as Pfaffenschlag, district Jindřichův Hradec, Mstěnice, district Třebíč (Nekuda 1975, 1985), etc. The same organization is also applied in the investigation of the extinct medieval village of Bystřec near Jedovnice, district Blansko.

*Jedovnice, EMV of Bystřec, district Blansko*

The archaeological investigation of the Moravian Museum in Brno in the area of the extinct medieval village (inhabited in the 13th century, abandoned probably at the turn of the 14th and the 15th century) has begun in 1975. The found documents illustrate the settlement process of the southern part of the Drahanská Highland, and provide information about the building development of dwellings, the material culture, the same as about crop and livestock production in the village, the social structure, the population environment, etc. (Belcredi-Nekuda 1983; Nekuda 1976; Belcredi 1986, 1986a).

A part of the investigation in 1984 to 1987 and in 1994 was also geophysical measurement carried out in the area of both the left and the right valley slopes of the Rakovec Brook, on four places of approximately 240 x 50 m, 200 x 50 m, and two zones of 50 x 50 m in a network of 2 x 2 m (Hašek et al. 1985a: 19-21; 1987: 25-28; 1988b; Bachratý-Hašek-Tomešek 1994). The objective of all those works was to locate the dislocation and sizes of the individual buildings, to compare the interpreted objects with the field sketch of the ground plan of the village according to (Černý 1970: 57, Fig.3; Černý – Černá 1964), and to include them into the situation found by the investigation.

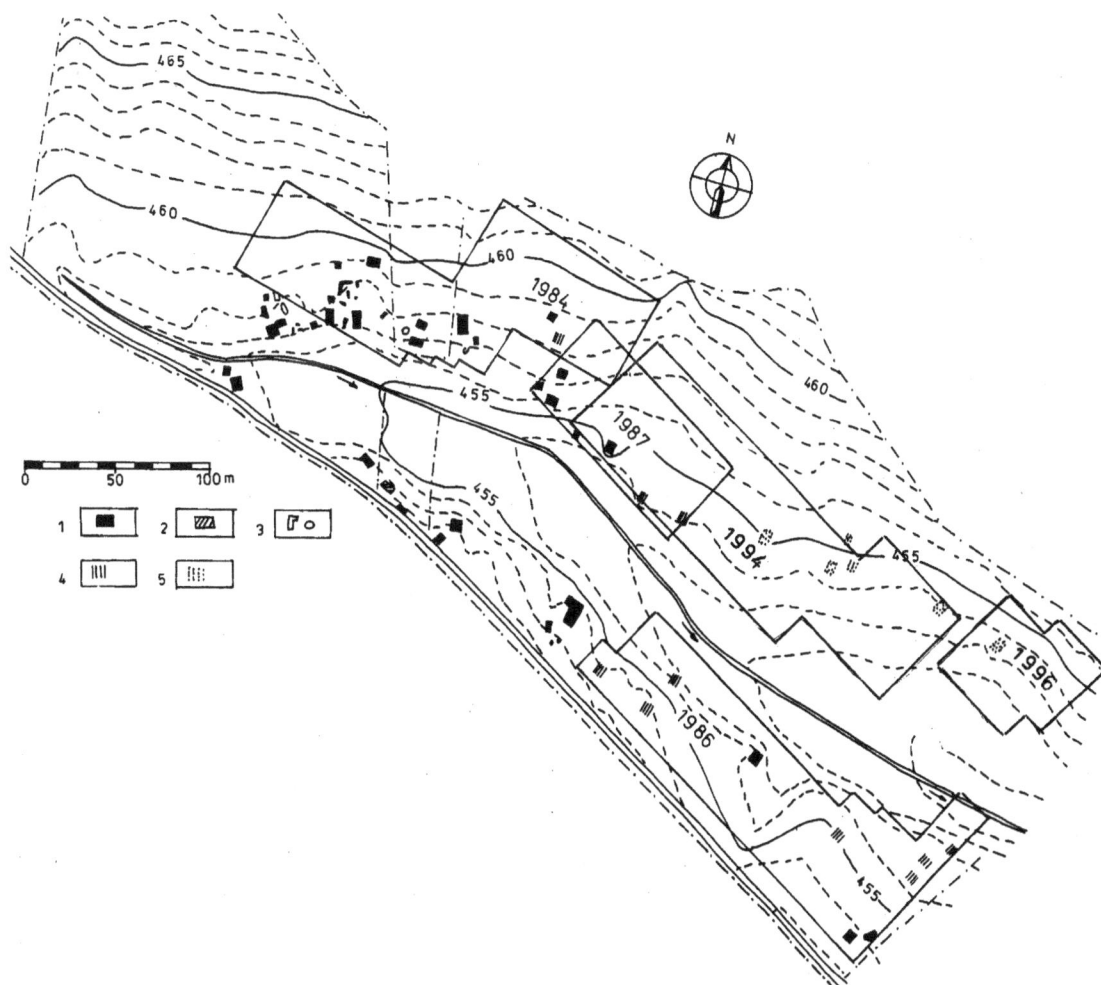

***Fig. 60.*** *Jedovnice, the extinct medieval village of Bystřec, district Blansko. Ground plan setting obtained by archaeological and geophysical research and gathering surface material (Černý 1970). 1- located settlement buildings, 2- destroyed buildings, 3- other objects, 4- buildings interpreted on grounds of geophysical measurement, 5- buildings assumed by Černý (1970).*

The processed data of magnetic properties of rock samples taken from the exposed objects of preceding investigation seasons (e.g. burned greywacke $\ll$æ = 26.5 . $10^{-3}$ u.SI, greywacke $\ll$æ = 0.1 . $10^{-3}$u.SI, etc.) proved that magnetometry is a suitable method for solving the above task.

The result of geophysical measurements are plans of $\Delta T$ and $T_z$ isanomales from which it is clear that the objects of investigation are prevailingly characterized by isometric anomalies of the magnetic field which indicate the position of the individual square and rectangular habitations (buildings and farming objects of interpreted dimensions of about 5 x 5 m to 10 x 10 m or even 15 x 10 m). Due to the intensity (as much as + 100 nT), the extent of the individual $\Delta T$ anomalies at the measured areas and from the data of measuring magnetic susceptibility in rock samples it can be assumed that due to the extinction of the village by fire their sources are chiefly the combinations of the burned layers (stone, daub), furnaces, Fe-objects, recessed habitation and production objects, etc. (Hašek-Měřínský 1987: 34-35; Hašek-Měřínský 1991: 96-98).

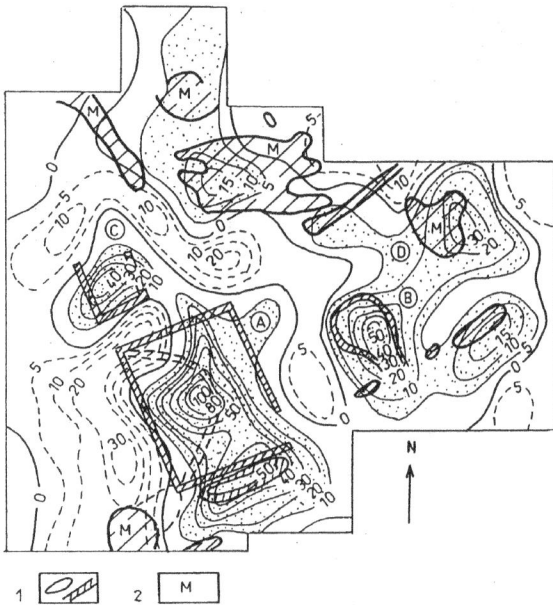

**Fig. 61.** *1- position of uncovered object 2- layer of plaster floor*

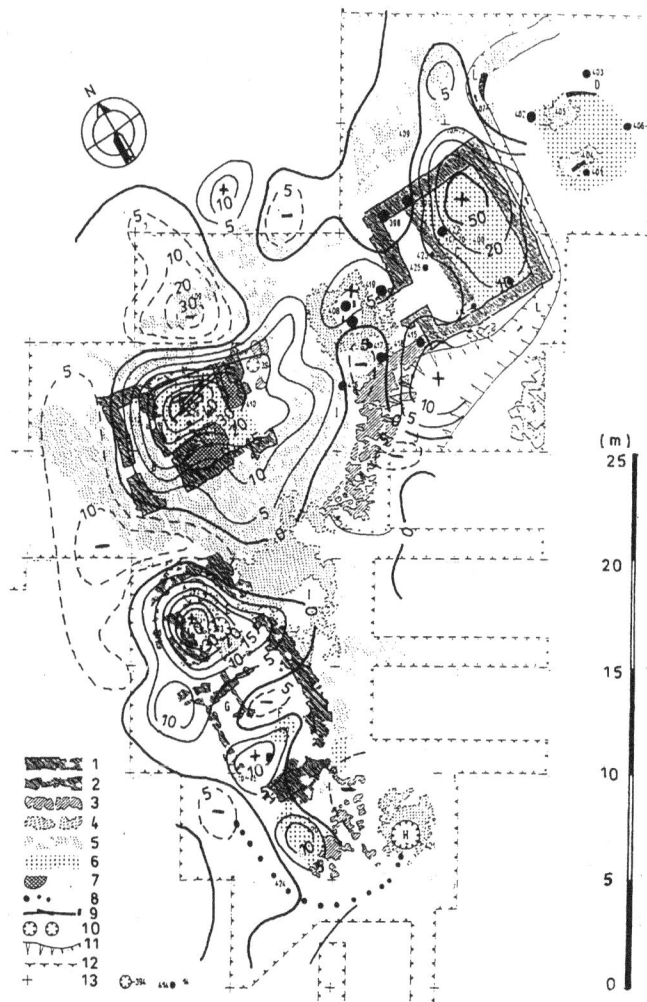

**Fig. 62.** *1- faced stone masonry, 2- unfaced stone masonry, 3- reinforcement of the slope with stones and/or flat stones, 4- paving, 5- areas filled with stones, 6- burned daubing, 7- area of fireplace, 8- pales, 9- remains of beams, 10- recessed objects, 11- field slope and/or edge of recess, 12- excavation edge, 13- intersections of the 5-m square net*

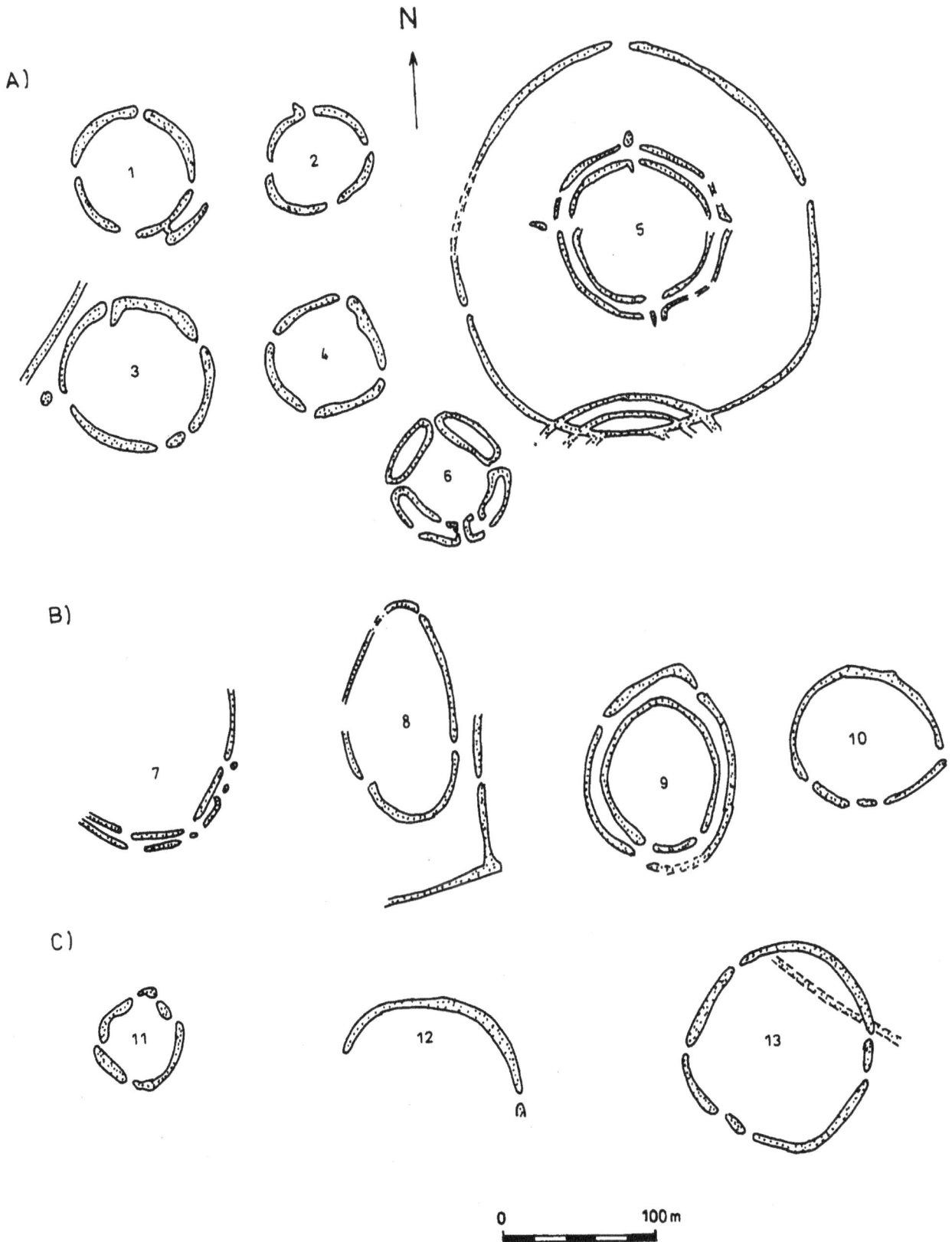

**Fig. 63.** *Circular structures from the period of Early Neolith, Bronze Age and structures from a so far unknown period. 1-Běhařovice, district Znojmo, 2- Němčičky, district Znojmo, 3- Rašovice, district Vyškov, 4- Vedrovice, district Znojmo, 5-Bolany, district Kutná Hora, 6- Lochenice, district Hradec Králové, 7- Šitbořice I, district Břeclav, 8- Šitbořice II, district Břeclav, 9- Šumice, district Znojmo, 10- Podolí, district Brno, 11- Troskotovice, district Znojmo, 12- Vážany nad Litavou, district Vyškov, 13- Vlasatice, district Břeclav.*

The comparison of the position of the individual habitations according to (Černý 1970: 57, Fig. 3) with the geophysical interpretation and the hitherto performed archaeological investigation (see Fig. 60) (Belcredi-Hašek - working diagram) it is possible to judge that the results of geomagnetic measurements rendered precisioning information about both the number of individual objects situated prevailingly on morphological elevations of the field and their overall sizes. An extent of settlement larger than that assumed on the basis of the surface investigation cannot be excluded.

Archaeological investigation carried on the western part of the area measured in 1984 (Fig. 61) confirmed the main results of the geophysical measurement.

An artificially raised terrace situated on the left bank of the brook at the western margin of the overall ground plan of the extinct village included foundations of the habitation denoted as No. V. The best preserved was space A covered with a burned layer of daub from the ceiling and maybe also from the walls with a large number of iron objects (axes, keys, tools, etc.). East of this object was investigated the first example of a separately situated granary whose lower recessed cellar part had walls provided with a stone screen (B). North of space A there was a smaller object with a light stone foundation, where also a burned layer of the ceiling daub was found (C). North of the granary a circular recessed object was exposed with a thick burned layer of sand, loam and stones (D) (Belcredi 1986: 54; Belcredi - Hašek - Unger 1990: 5-23; Hašek - Měřínský 1991: 98-99).

A similar situation is also found in places of the right bank of the brook near the edge of the present wood near the southeastern margin of the whole ground plan of the extinct village (see Fig. 62).

The results of the 1989-90 investigation proved the foundations of three buildings (Fig. 62). The best preserved was the northeastern space with a residential building (A,B) and a chamber (C) covered with a burned layer of daub probably from its ceiling, etc. In the southeastern direction of these objects also farming buildings (E,F,G) were exposed (Hašek-Kovárník 1996).

## 4.2. Investigation of fortified formations

The objective of archaeogeophysical prospection of the vallum bodies (banks, walls with wooden or stone structure, stone walls), moats, palisade rings, etc. is to clarify their position (particularly in cases when, by a long-term soil cultivation and further interventions their course was destroyed), the structure, ground plan layout, dimensions, places of entrances and gates, bastions, tower objects and other building elements, including the determination of the overall ground plan, extent of the fortified area and further settlement also outside the fortified areas.

The processing of physical properties of samples from the individual exposed objects indicated the possibility of employing both magnetometry and geoelectric methods in solving the tasks of the above issues.

### 4.2.1. Prehistoric Circular Structures

In recent years aerial prospection has begun to be used on a large scale in combination with surface geophysical, particularly magnetometric measurements. The two methods complement each other. Whereas aerial prospection follows the traces of archaeological objects visible on the surface of the field and transfers their location into maps, geophysical works mean a further, qualitatively different stage which can provide a more detailed overview of the cumulation of objects at the station and in a number of cases, such as the circular structures (Neolithic, Bronze Age) - roundels and the ground plan layout and dimensions, exactly situated into the map. On the above areas the archaeological surface investigation a reconnaissance investigation which forms the third stage, essential for the verification of the function and chronological position of the objects discovered by aerial and geophysical prospection as well as the credibility and the rate of accuracy of the ground plan diagrams of the studied stations is then carried out and (Bálek-Hašek-Měřínský-Segeth 1986: 571; Bálek-Hašek-Ondruš-Segeth 1989: 5-16; Bálek-Hašek 1996: 7-20).

Simple circular structures (flattened circle, wide oval, etc.) dating back to Late Neolithic and Early Bronze Age found in south and southwest Moravia and in Bohemia by means of the two prospection methods are in essence of two kinds. They are either small simple or double formations of the diameter up to 70 m - Běhařovice, Němčičky, district Znojmo, Lochenice, district Hradec Králové -, or objects of medium size, diameter of 100 m and more - Vedrovice, district Znojmo, Rašovice, district Vyškov, Hluboké Mašůvky, Šitbořice, district Břeclav, Šumice, district Znojmo, etc. This group can possibly include hitherto archaeologically undated simple, double and triple circular and semicircular structures near Vážany nad Litavou, district Vyškov, Troskotovice, Vlasatice, district Znojmo and Horákov, district Brno (Hašek-Kovárník 1996a). The ground plans of all those objects according to the geophysical measurement are given in Fig. 63.

*Straubing, Lower Bavaria (Germany)*

The objective of geophysical works carried out in 1992 (Engelhardt-Hašek-Unger 1994) at the polycultural settlement (Neolithic - La Tène) in a wide surrounding of the hamlet of Lerchenhaid west or northwest of Straubing, on the river route of the Danube at an area of about 5 ha, sloping mildly towards the south and near the road E 5 (Regensburg - Landau a.d. Isar), was to track the course and to determine the overall ground plan situation of two Neolithic moats of the Linear Pottery Culture, partly exposed by the investigation carried out by the Bayerisches Landesamt für Denkmalpflege Landshut in 1987 and to locate further recessed settlement and other objects.

In order to solve the outlined tasks, at 20 sectors of the basic size of 50 x 50 m the magnetometric method was applied. The measurement was performed in a regular network of profiles and points of 2 x 1 m or 2 x 2 m.

From the systematic processing of the results in the form of a correlation diagram (Fig. 64), a number of anomalous elements whose sources are mostly different recessed objects at the archaeological station studied can be specified.

Among the predominant ones can be considered three relatively extensive linear $\Delta T$ anomalies (I, II, III) arranged into a simple circle or oval, with different size, width and orientation.

In the first of them (+ 6 nT, I) which could not be entirely locatad by geophysical measurement due to a large built-up area in the northwestern part and disappearance of its further course in the southern part, it is assumed to be of oval shape, with the dimension of the long axis of about 300 m, oriented

in the direction NW - SE, the short axis being about 200 m. The width of the anomalous belt, which mya in places be also doubled (e.g. the western margin), is about 2 m, the depth of the source 1.5 m. The positions of 4-5 entrances into the supposed moat-surrounded formation can be situated approximately into the direction of the cardinal points. Relatively more marked in the magnetic picture is the course of the second structure (+9 nT, II) of approximately circular ground plan with a diameter of 185 m. The simple circle is interrupted, as in the first case, by four entrances into the area, each about 3 m wide. The possible position of two further partial entrances near its southern margin cannot be excluded. The width of the specified moat is about 3.5 m, the

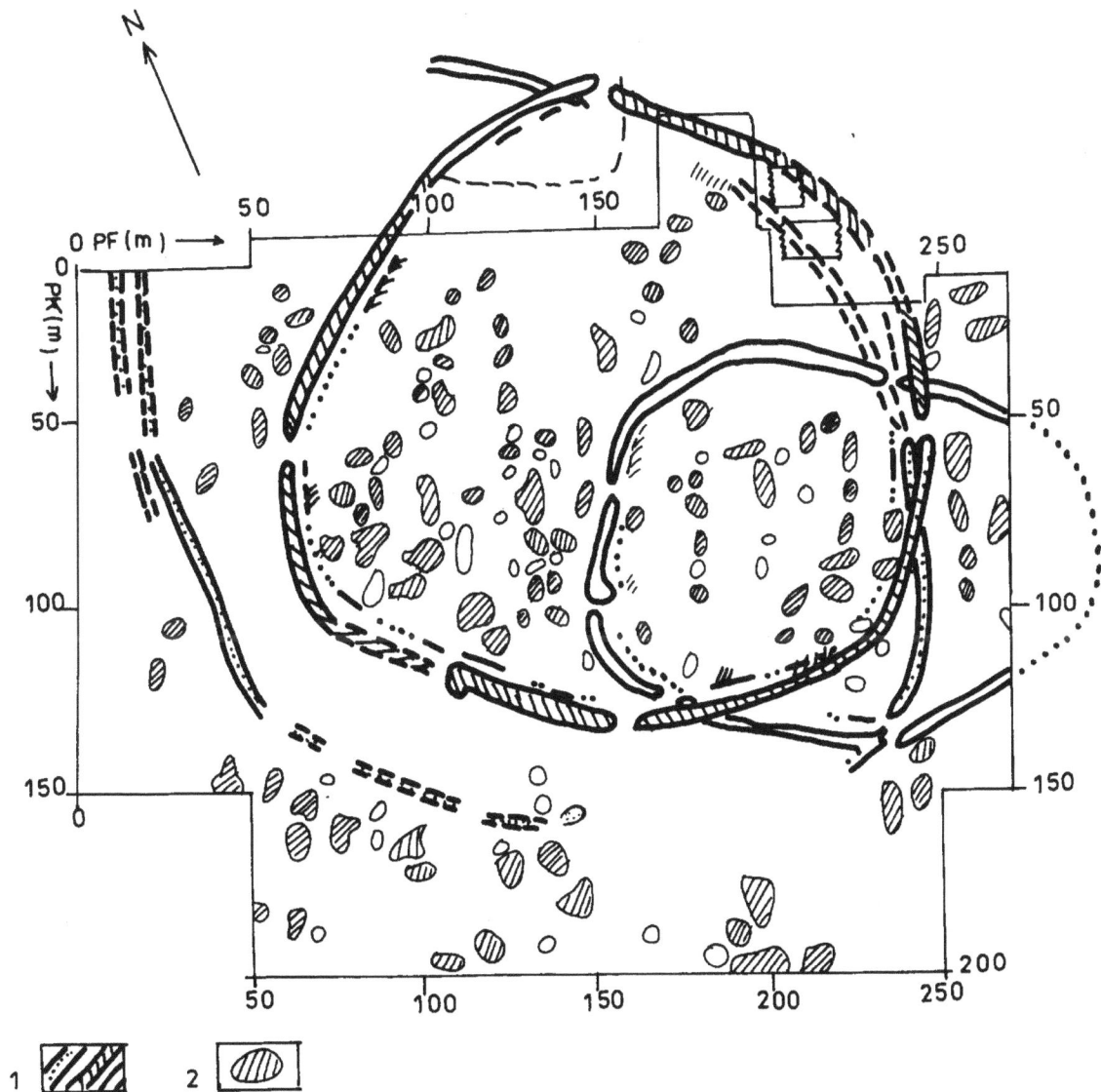

*Fig. 64. 1- ditch alignment, 2- three-dimensional sunken object*

depth max. 2 m. From the results of the measurement it is possible to assume the course of relics of the doubled palisade ring which runs parallel to the ring from the inner side at the distance of about 5 m. Among the most apparent can be included the third structure of oval shape (+12 nT, III) situated in the eastern part of the area studied. Its long axis, oriented approximately in the E - W direction is about 150 m long, the short one about 110 m. It is assumed that this formation, although smallest as to the areas, marks the presence of a moat about 3 m wide, about 2 - 3 m deep. Unlike the above objects, the orientation of about 5-6 simple entrances cannot be proved unambiguously in the direction of the cardinal points.

Inside all three supposed structures delimited by moats was a large number of prevailingly isometric positive $\Delta T$ anomalies of different surface area, from 5 x 5 up to about 10 x 10 m (see Fig. 64). Their sources are probably positions of recessed habitation and other objects of different ground plan, filled with a cultural layer of differential magnetization towards the surrounding environment. According to geophysical data the largest concentration of settlement is assumed to be inside the second and the third structure, or at the southern side of the measured area, outside the first specified formation.

Linking up with the data of the performed archaeological investigation we think that the most extensive oval structure delimited by the moat, oriented in the NW - SE direction, can be assigned to the oldest exposed moat dating back to Early Neolithic (Linear Pottery Culture), the approximately circular formation to a later structure of the same period of the Neolithic and for the oval moat formation with the long axis in the approximately E - W direction an even later time period cannot be excluded. In the first two structures delimited by moats it is possible to notice 4 to 6 simple entrances about 3 m wide, with the geographic orientation, i.e. in the direction of the cardinal points (bearing + 12°). This assumption cannot, however, be confirmed in the third structure (see Fig. 64). Remains of a palisade ring can be assumed to be at the inner side of the second, and maybe also the third moat.

The area of interest was densely inhabited. Early Neolithic and later settlement was proved by the investigation of the locality. It can involve settlements and farming buildings, cultural pits, loam pits, etc. The results of geophysical works demonstrated the largest concentration of recessed objects especially inside the fortified areas. To the W or SW their density gradually decreases.

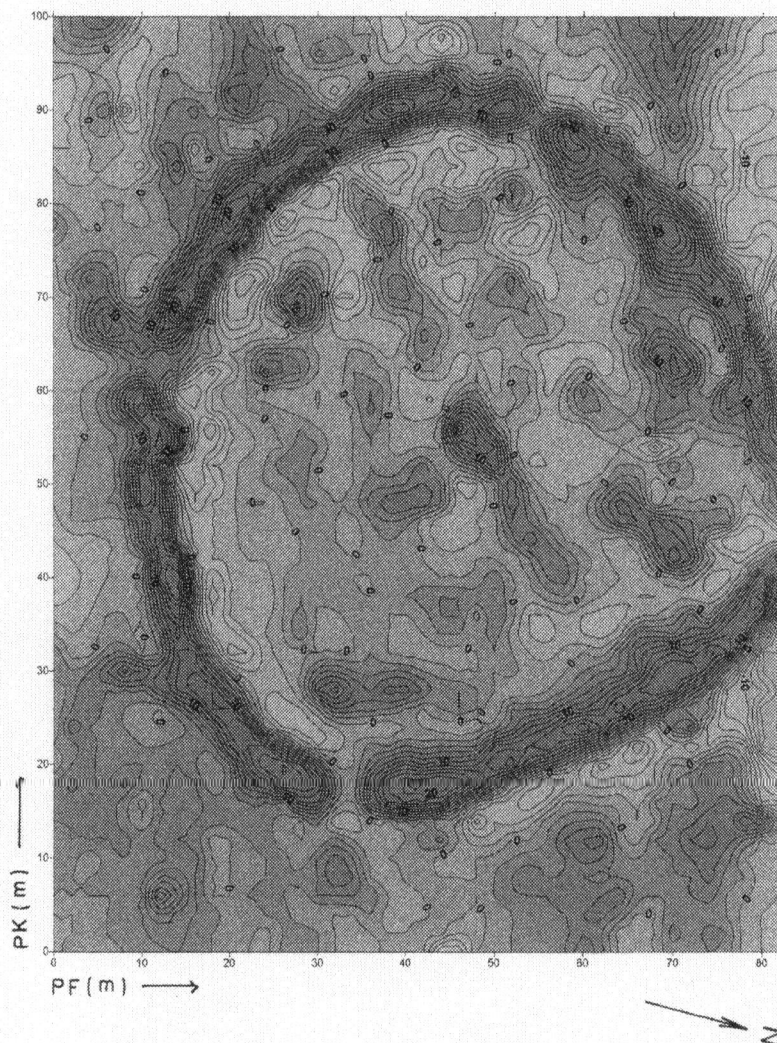

**Fig. 65.** *Vedrovice, district Znojmo. Map of surface-plotted $\Delta T$ values from the area of neolithic circular structure.*

60

*Vedrovice, district Znojmo*

Among the numerous localities of the central European Neolithic, where settlements with the Linear, Stroked and Moravian Painted Pottery Cultures are documented, the attention in recent years has been paid particularly to Vedrovice in the Bobravská Highland, about 40 km SW of Brno. The most significant part there is the representation of the settlement with Linear Pottery Culture (6th millenium B.C.) which, on the basis of surface finds, can be tracked down to an area of about 14 ha. This settlement became the focus of a systematic field investigation carried out since 1961 by the Moravian Museum in Brno (Ondruš 1966). After deep ploughing of the greater part of the inhabited area in 1982 suggestions of a moat were found encircling a settlement of the Linear Pottery Culture. Its existence was confirmed by archaeological probes. The finds date the moat to the period of the second settlement of the village, i.e. to the middle stage of the Linear Pottery Culture (Bálek-Hašek-Ondruš- Segeth 1989: 6-13).

In the eastern part of the habitation, in places where deep ploughing had not been done and where the settlement by the Linear Pottery Culture is overlapped by the further Neolithic Moravian Painted Pottery Culture, aerial prospection registered a hint of a circular structure. The structure which was assumed to be a Neolithic circular object delimited by a moat on the basis of a measurement diagram (Bálek 1985: 113-114) and by direct observation (Kovárník 1985: 104) according to typical signs is situated in a close proximity of the road not far from the southwestern margin of the community of Vedrovice.

By surface magnetometric measurement carried out in a network of 2 x 2 m with a densification to 1 x 1 m, processed into a map of ΔT isanomales (Fig. 65) and the spatial image (Fig. 66) at the area of 80 x 100 m a regular simple circular object was found with the outer diameter of about 75 m and the moat depth of 6 - 8 m, interrupted by four entrances 2.5 to 4 m wide, situated at the axes with a deviation from the present direction of the cardinal points of about 8°.

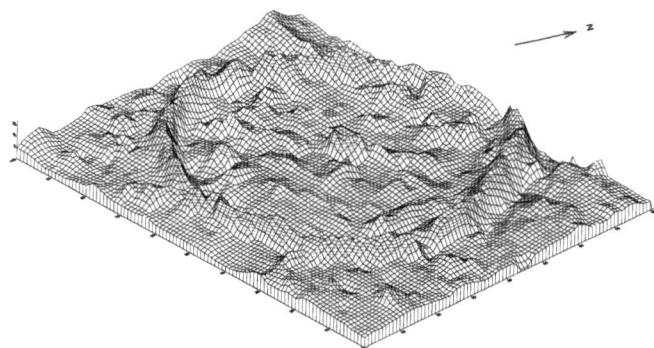

*Fig. 66. Vedrovice, district Znojmo. Three-dimensional plot of ΔT anomalies.*

The depth of the moat probably varies only very little according to the size of the ΔT values, whereas the width varies considerably. The maximum ΔT anomaly is approximately identical with the axis of the moat. A small outer deviation of the moat was found at the southeastern side of the object. The largest width (maybe the doubling of

the moat of the Linear Pottery Culture and with the Moravian Painted Culture) is at the northeastern part. In the northwestern quadrant at the inner side of the moat a large object (probably a loam pit) can be assumed.

Conspicuous is the singling out of isometric anomalies ΔT in the proximity of the entrances into the object and the existence of a small linear anomaly in the direction of the axis NE - SW in which a sign of a small moat cannot be excluded. The whole area of interest is magnetically considerably impaired prevailingly by recessed objects of the settlement and production character, interpreted both inside, but also outside the moat (Bálek - Hašek - Měřínský - Segeth 1986: 566-569, Figs. 6-7; Bálek-Hašek-Ondruš-Segeth 1987: 141-153; Hašek-Měřínský 1991: 109-110; Bálek-Hašek 1991: 33-37).

From aerial prospection another extensive structure was observed to the southwest and northwest of the above circular moat. By magnetic measurement in the studied area an extensive object was found of elliptic ground plan, interpreted as a moat structure of the width of about 5 m and the depth of 1 - 1.5 m (Fig. 67) (Bálek - Hašek - Měřínský - Segeth 1986: 568, Fig. 8).

The moat itself runs in the northeastern sector of the studied area in the inner eastern part of the circular object and it turns towards the southwestern or western side. Its situation is interrupted in several places on the one hand by later loam pits, dimensions about 20 x 10 m (the southeastern part of the studied area), and on the other hand by entrances into the settlement. Conspicuous is also a small local arching of isanomales ΔT at the inner side of the moat, which could be due to the location of furnaces.

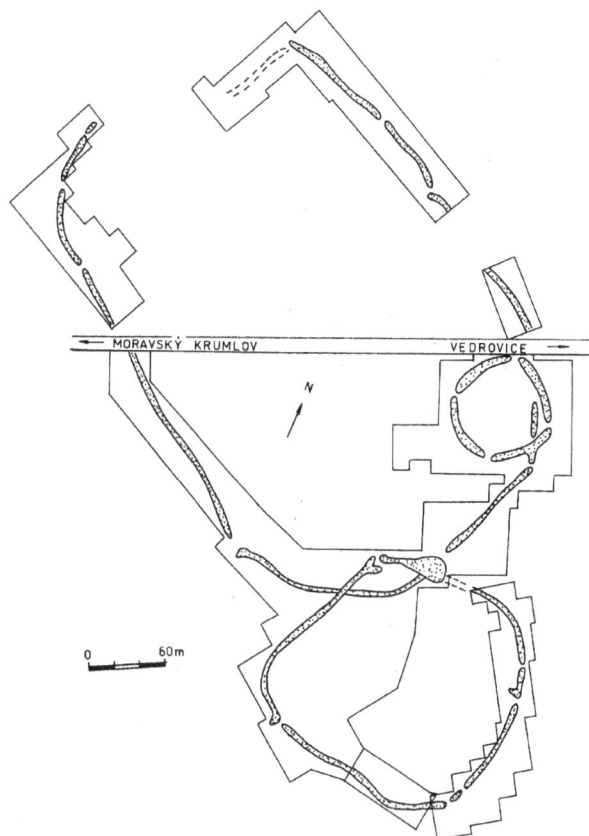

*Fig. 67. Vedrovice, district Znojmo. Scheme of ditches.*

In the southernmost sector another structure of approximately oval shape was localized by geophysical measurement. Its long semiaxis is about 170 m long, the short one about 150 m (see Fig. 67) (Hašek - Měřínský 1989: 103-151; Hašek - Měřínský 1991: 111). From the processing four entrances are assumed, oriented in the NW - SE and NE - SW directions, respectively, with maximum width of about 4 to 6 m. The course of the moat is geophysically less conspicuous on its southern side due to its location on an exposed slope affected by extensive soil erosion. A large cumulation of habitation inside this object was not determined by magnetic measurement.

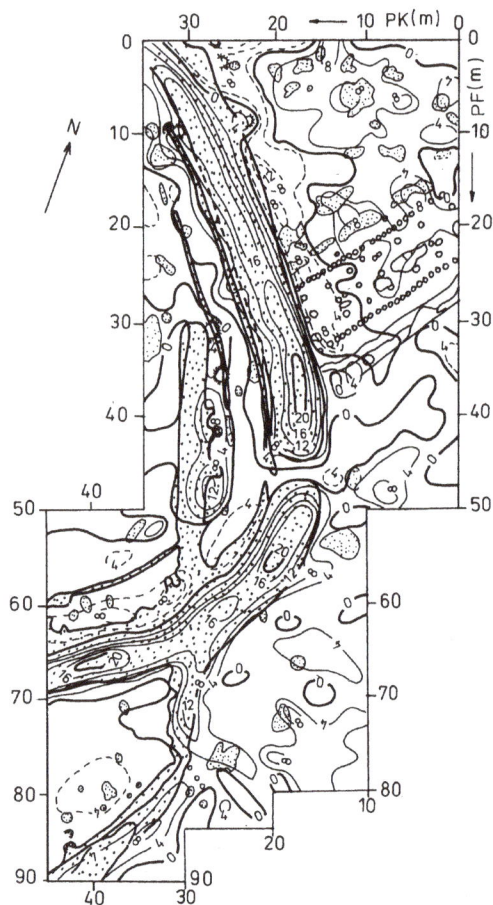

*Fig. 68. Vedrovice, district Znojmo. Results of the archaeological research compared with a section of the map of ΔT isanomales.*

Archaeological investigation started in 1985 was concentrated particularly on the eastern sector of the measured area (Fig. 68). So far the northeastern quadrant of the circular moat has been exposed, recessed into a thick layer of calcareous loesses in the length of 85 m which can be approximately dated into the earlier stage of the Moravian Painted Pottery Culture (æ = 0.61 - 0.92 . 10⁻³ u.SI). At the same time, in the length of 60 m a moat of the Linear Pottery Culture was followed (æ = 0.36 - 0.59 . 10⁻³ u.SI) at the outer side of which furnaces were located (æ = 0.55 - 0.77 . 10⁻³ u.SI). By its shape the moat corresponds fully to both the results of geophysical measurement and the information obtained from the investigation of this moat in the western part of the settlement (Bálek-Hašek-Ondruš-Segeth 1987: 141-153).

The filling of the two above moats consists of a compact dark brown layer of rainwash material. The above values measured in the axis of the two moats drop in the direction of greater depths.

By probing works carried out in the southern part of the area studied in places of the interpreted position of the third structure of a circular ground plan (see Fig. 67) a moat with triangular cross section was discovered with the depth and width of about 3 - 3.5 m which has not been archaeologically exactly dated so far.

The measured local anomalies ΔT in the exposed area are due to small recessed objects of habitation character. Neither an overground object at the outer margin of the MPC moat nor the double palisade ring at its inner side have been found by geophysical works.

Using aerial prospection combined with geophysical measurement on the surface and archaeological surface investigations in the process of finding and recognizing the issues of circular Neolithic objects in Moravia and in Bohemia their expediency was found, as well as their accuracy, both for the amount of scientific information and for saving time and, last but not least, also financial costs.

### 4.2.2. Prehistoric Fortified Settlements

*Kokory, district Přerov*

East of the village of Kokory, in the position called "Hradisko" (ground elevation 252.7 m), above the river Olešnice remains of a prehistoric hill-fort (the Věteřov Culture, Hallstatt) were discovered situated on an area of about 0.7 ha. At its southern sector an arch-shaped fortification has been preserved consisting of an earth bank, a moat and a small defence line - vallum at the outer side of the moat. The earth bank is about 0.8 m high, the height difference between its crown and the moat bottom is about 3 - 4 m. The defence line - vallum in front of the moat is about 1 m high. The northern slope of the acropolis is damaged due to surface extraction of stone.

By a finding probe carried out across the earth bank, the moat and the small defence line - vallum (21 x 1 m) by the Museum of Local History and Geography at Olomouc in 1986 (Dohnal 1988) near the SW margin of the station it was demonstrated that the original settlement was probably not fortified. That was proved by the discovery of a buried cultural layer about 25 cm thick below the body of the earth bank. On the inner side of the earth bank its thickness was up to 50 cm. The structure of the earth bank body was loam-and-stone. It contains charred pieces of wood which probably held the front side of the earth bank. The overall wall width was about 1.5 - 1.7 m. In front of it a berm about 2 m wide was found, also covered with a brown culture layer. The moat was dug into weathered Culm rocks. Its depth and shape was, however, impossible to establish by the investigation. Its depth does not seem to exceed 1 m. The defence line - vallum in front of the moat is not layered.

The objective of geophysical works carried out in 1995 (Hašek-Tomešek 1995) was to make up a geophysico-

geodetic plan of the station at the area of 130 x 90 m and to
map the situation of settlement inside the hill-fort.

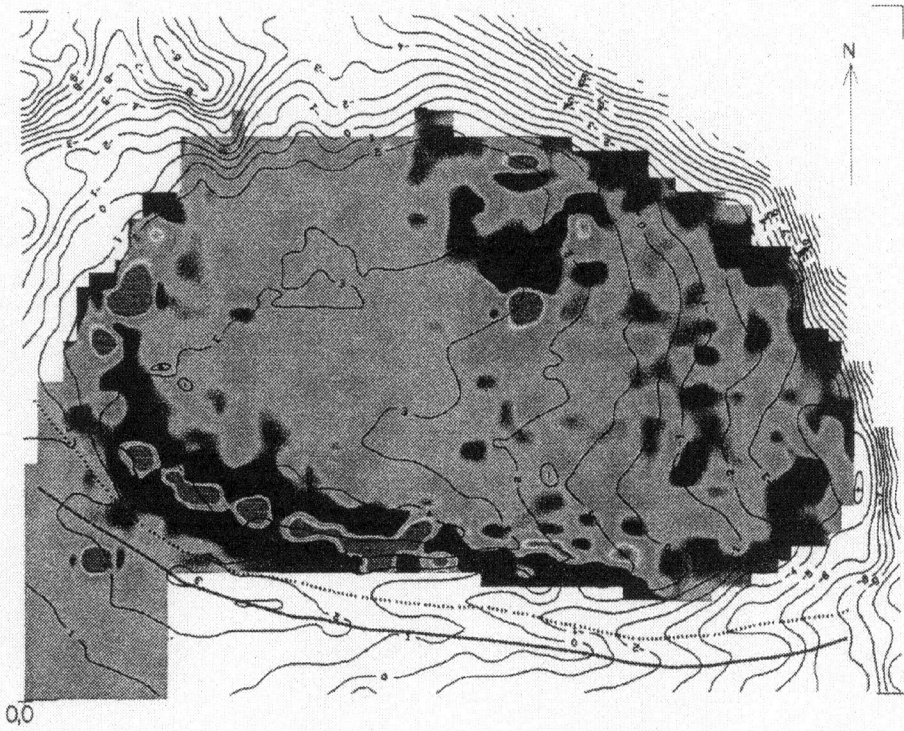

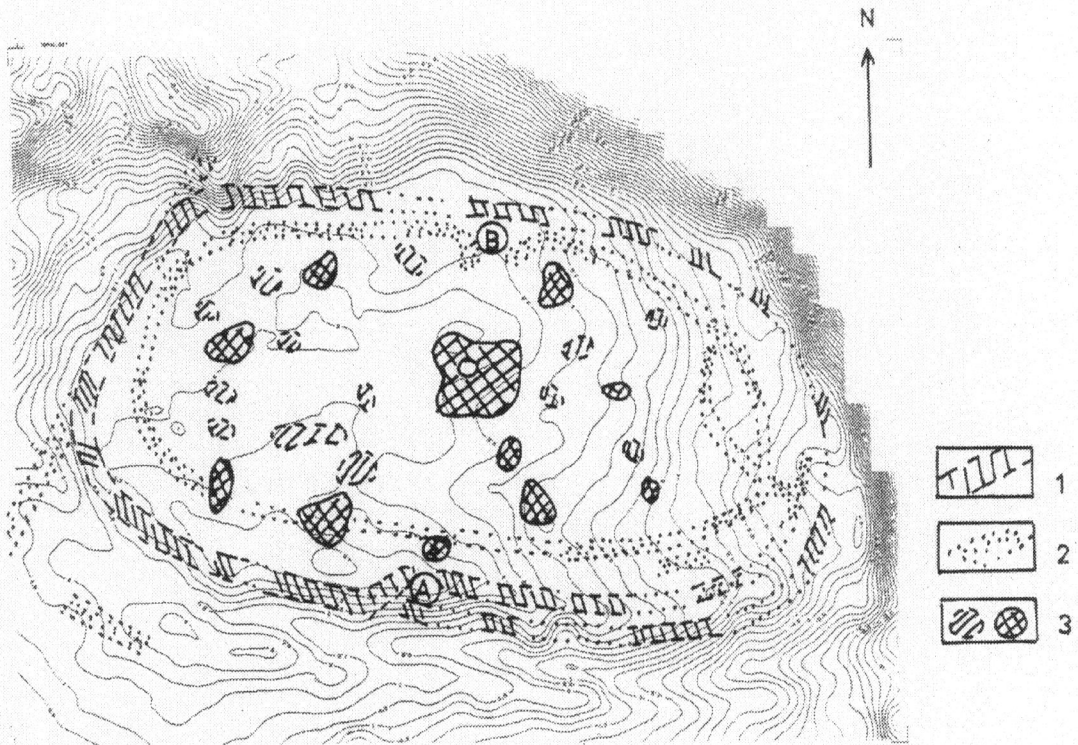

***Fig. 69.*** *1- dicht alignment, 2- position of older ditch, 3- sunken object*

The compiled geodetic model of the acropolis (Fig. 69) demonstrated the existence of a fortification system (the earth bank, the moat, the defence line - vallum) only in the southern and the southeastern sectors of the station, a morphological plateau in the central and the western parts and probably recent terrain adaptations at the eastern and the northern sides which may have caused a certain absence of expected archaeological objects. Along the northeastern and the western circumference of the hill-fort there is a steep drop of the terrain relief in the direction to the valley of the Olešnice river.

From the results of magnetic measurement by the method of vertical gradients (see Fig. 69) several positive linear and isometric anomalies $T_z$ were found whose sources may be both the cultural layer from the overground and recessed settlement and other objects of different ground plans and also burned layers of loams and clays and/or charred pieces of wood from the structure of the earth bank body. The sizes of the interpreted structures vary from about 4 x 4 m to 25 x 5 m. Characteristic is their linear orientation along the circumference of the studied area and a more or less regular inner layout.

Among the most conspicuous can be included positive anomalies $T_z$ at the outer side of the earth bank in the southern and southeastern parts of the hill-fort which further turn to the northern direction (see Fig. 69). Thus they delimit the area of interest in the planar section along its whole circumference. A reduced correlation of those magnitudes can only be observed on the northern and the northeastern edges of the structure, where they could possibly have been obliterated due to later terrain adaptations, extraction of stone, etc. From the above, in comparison with the results of the finding probing it can be deduced that the linearly oriented anomalies of the magnetic field are, with respect to their width, due to the burned positions of loams in combination with charred pieces of wood from the wall structure and with a possible extension of the cultural layer into the space of the berm. The possible doubling of the wall cannot be excluded either, or a change in its structure at the southeastern edge of the ringwall in the length of about 75 m and a possible entrance (gate) - denoted as A (see Fig. 69) of the width of about 3 - 4 m. At the southwestern sector of the structure the course of a destroyed road is assumed. The overall distribution of anomalies $T_z$ hints the possible fortification of the whole area, though probably with a different character of its structure.

About 5 - 7 m from the crown of the defence line - vallum towards its inner side, less intense linear anomalies $T_z$ were measured of a more local character, bordering the above zone of intense anomalies of the magnetic field along its whole circumference. With respect to the overall layout (see Fig. 69) either a course of some narrow and shallow (older?) fortification (moat?) can be interpreted there or the position of a cultural layer from recessed objects in the proximity of the present defence line - vallum body. In the case of the former variant the assumed entrance could be situated in places denoted as B (see Fig. 69).

The source of positive isometric anomalies $T_z$ inside the studied structure might be - according to the conspicuousness - a thick cultural layer from the filling of objects of max.

ground plan of 5 x 5 m. Interesting is a relatively extensive anomaly $T_z$ detailed at the approximately highest point of the field (see Fig. 69). The possible occurrence of recent Fe-waste is assumed there, also in combination with a major archaeological object. Anomalies $T_z$ at the outer side of a smaller earth bank can similarly be formed either by the position of the cultural layer or even by recent bodies.

The above assumptions will become the object of further investigation.

*Mušov, district Břeclav*

A document of the Roman military and building activities of the time from the beginning of the first century A.D. up to the Marcoman wars in the region north of the Danube is the military fortified camp, situated NW of the former community of Mušov, about 17 km from Mikulov, on a mild elevation called "Burgstall" (today "Hradisko") which has an outstanding strategic position above important crossroads of ancient routeways at the confluence of the Jihlava with the Svratka and the Svratka with the Dyje.

On a flat knoll there are the only safely documented Antique buildings in the Czech Lands. In 1926-28 remains of two masonry buildings were exposed there, foundations of a rectangular habitation building with four rooms and a part of the western wing of another complex of buildings serving as a bath.

The systematic investigation of the AI CSAS in Brno (now the AI AS CR Brno) have since 1985 yielded some new views of the issue of the Mušov buildings (Tejral 1986: 395-410; Tejral in Podborský et al. 1993: 440-443). Above all it has been found that the whole knoll of the area of about 9 - 10 ha is surrounded by mighty fortification consisting of an almost 2 m deep moat and a strong wood-loam bank wall. The Roman age is documented by numerous objects excavated from the moat (fragments of pottery, bricks with the seal of the Xth legion, etc.).

The task of geophysical works carried out on the one hand at the knoll of "Hradisko" and further at its eastern sector was to verify the assumed course of the fortification and/or further objects found by aerial prospection.

From the resulting processing of the measured data in the inner space of the camp (Fig. 70) it followed that the overall course of the magnetic field was interrupted by a number of anomalies $\Delta T$ of local character, whose causes may be both archaeological objects, and also large recent Fe-objects coming from the conclusion of World War II (the strategic position is also stressed by the fact that in the spring of 1945 the German Wehrmacht opposed arduously the proceeding Red Army there).

Much more interesting is a linear anomaly $\Delta T$ in the NE and E sector of the measured area (+20 nT) turning into an oval which can be followed from profile 0 of picket of 15 m up to profile 5 picket of 50 m an further from profile 80 picket of 0 m up to profile 20 picket of 30 m which corresponds, as found by the investigation of the course of the moat, probably from the time of the Únětice Culture of the Early Bronze Age (1900 - 1500 B.C.) in the width of about 4 - 5 m (see Fig. 70). Its further continuation in the southeastern direction cannot be found due to the position of a vineyard with an Fe-structure which screens the effect of the object of investigation. Similarly, also relatively extensive anomalies - linear structures at profiles 60 to 110 picket - 10 m (+ 8 nT),

in the surroundings of profile 64 picket 0 - 50 m (+ 8 nT), profiles 0 - 30 picket - 30 m (+ 20 nT) can be in relation to the fortification of the military station. In the other separated anomalies $\Delta T$ (see Fig. 70) it can be above all the effect of recessed and other objects.

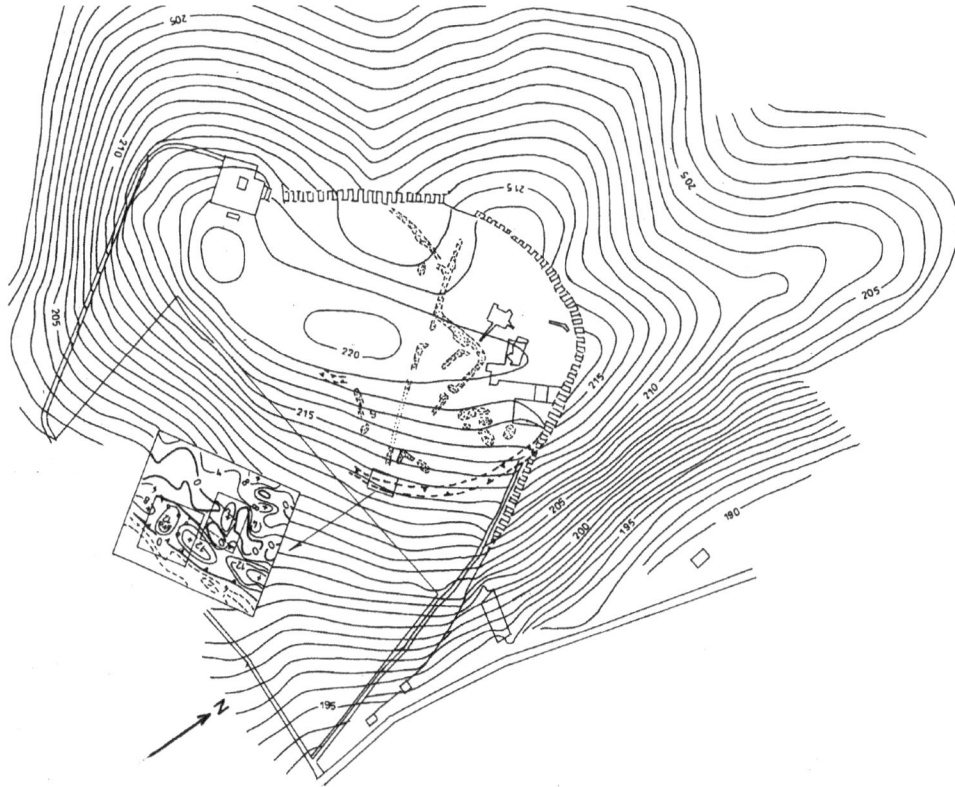

**Fig. 70.** Mušov, district Břeclav, "Hradisko". Map of $\Delta T$ isanomales and the situation of the excavated objects.

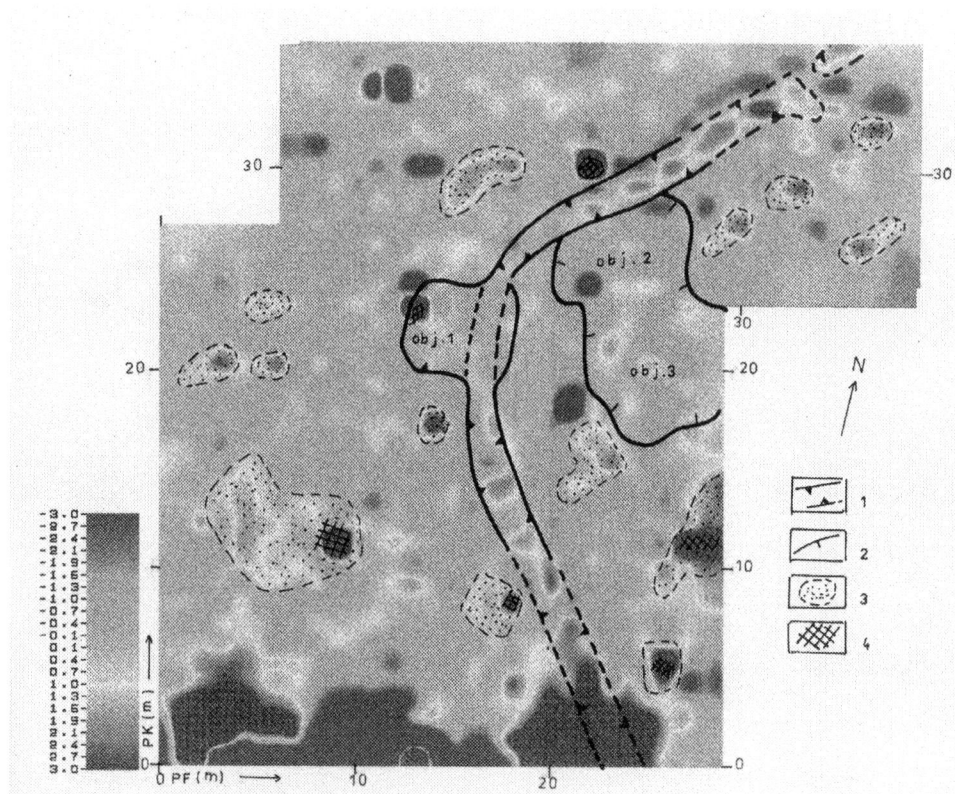

**Fig. 71.** 1- ditch alignment, 2- uncovered sunken object, 3- object resulting from interpretation, 4- intensive $T_z$ anomaly (Fe-object, furnace)

Archaeogeophysical prospection carried out about 2.5 km SE of the military area Mušov I - "Hradisko" in the area called "Na pískách" was to verify the existence of the northeastern corner of the Roman field or marching camp, in places overlapped by Germanic settlement objects, located from aerial photographs (Kovárník 1993: 108-110) for the subsequent archaeological investigation.

From the map of grad. $T_z$ (Fig. 71) a linear anomaly of the magnetic field is interesting, interpreted as the position of a moat 2 - 2.5 m wide whose axis turns from the direction SSE - NNW to ENE - WSW and further a number of three-dimensional anomalies $T_z$, bordering the above zone which can picture the ground plans of recessed objects or perhaps even furnaces. A small impairment of the course of the moat can be assumed at the northern margin of the studied area. At the inner side of the separated linear structure, according to the measured negative anomalies $T_z$, the possibility of the existence of a small gravel bank cannot be excluded.

The archaeological investigation confirmed the results of geophysical processing. Besides the located moat, in places of isometric anomalies $T_z$ (see Fig. 71) three major recessed objects were discovered dating back to the 2nd - 3rd centuries A.D.

### 4.2.2. Slavonic Fortifications

*Břeclav - Pohansko, district Břeclav*

Long-term systematic investigations of the Great Moravian ringwall "Pohansko" has been carried out by the Institute of Archaeology and Museology, Faculty of Arts, Masaryk University, Brno. Since 1959 several burial grounds have been investigated there, one of them a cremation burial ground, foundations of a church building, the area of a magnate farmstead, a number of settlement groupings and the structure of a defence line - vallum fortification (Fig. 72). Extensive settlement agglomerations were also discovered in the form of suburbs, particularly to the southwest and to the northeast of the ringwall, where in the case of the former, in the years 1975-79, in connection with the planned adaptation of the river Dyje, an extensive rescue investigation was going on (Dostál 1984: 141-142).

In 1979, in the eastern part of the ringwall, in places of a marked depression in the body of the fortification defence line - vallum, a detailed geophysical measurement was carried out (Hašek - Mayer - Pantl et al. 1980: 26-30). In that space the existence of a gate could be assumed leading from the east to the ringwall. By magnetometric measurement on the area of about 20 x 60 m in the network of 1 x 1 m intense anomalies $\Delta T$ (+ 280 nT) were found at the margins and in the center of the above depression (Dostál - Hašek - Měřínský - Vignatiová 1981: 55-57) (Fig. 73).

These anomalies were evoked, as also shown in earlier works and subsequent investigation following them at other sectors of the ringwall, by layers of burned loam. In geoelectric measurement by the SRP method (A2M1N2B m) increased specific resistances were due to layers of stones. Archaeological investigation in the measured sector was started in 1982 on an area of 10 x 30 m. In the space of the

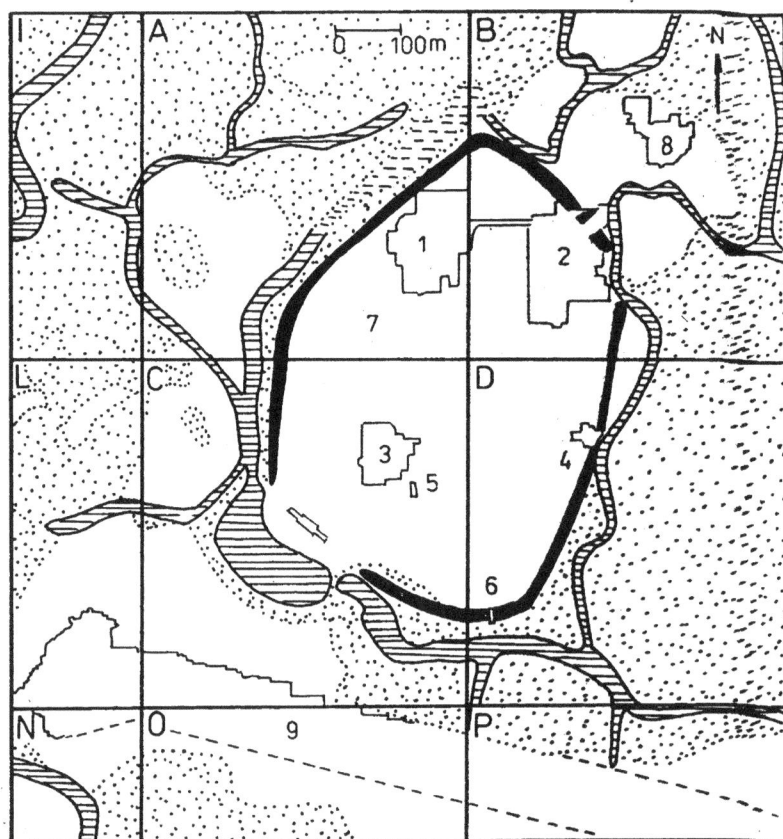

***Fig. 72.*** *Břeclav-Pohansko, district Břeclav. Situation of the archaeological research and the areas of geophysical prospection.*

section a gateway of the front type was discovered and studied. It is delimited by groups of four posts securing the two sides, reinforced by thick planks. The clearance of the gate was 245 cm, the overall width, together with the structure of the walls was 325 cm and the length of the passage 635 cm. At the outer side the structure of the gate walls protruded about 50 cm before the face of the wall. The entrance was probably closed by a gate, because at the outer entrance an iron hinge was found and in the innner entrance a part of a massive iron locking bolt. Thick layers of burned loam in the space of the gateway, differing markedly in resistance and magnetism from the grayish brown material of the defence line - vallum body, arose from the destruction of the gateway by fire. The gate, wooden walls and evidently also of the tower-like superstructure above it were burnt. Also the further course of the field investigation outside the space of the gateway confirmed the preliminary conclusions obtained on the basis of geophysical works. In places where there are no more anomalies of the magnetic field, the inner part of the gateway was continued by a road leading into the ringwall. It was not fixed, its course is delimited by a group of regularly laid out and uniformly oriented graves respecting it from the southern side (Dostál 1984; Dostál - Hašek - Měřínský - Vignatiová 1981: 55-57). A more complicated situation appeared at the outer side of the gateway, where archaeological investigation found a 15 cm thick layer of blue clay with charred pieces of wood. On that layer and sometimes also inside it stones were found from the destruction of the front wall of the Great Moravian fortification (Štelcl-Dostál 1984). The layer of blue clay was sinking roughly from the level of the surface to the depth of 1 m. On the basis of this information an assumption arose that at the time of Great Moravia close in front of the gateway there was a river bed which completed the fortification system of the ringwall. By geoelectric measurement of 1984 by the VES and SRP methods (Voňka 1985: 24-26) it was possible to locate a river bed characterized by high specific resistances (dry sands near the surface) in the proximity of the defence line - vallum. Pedological probes found that they continued partly even under the defence line - vallum besides the present arm of the river Dyje. In the space of the gateway the above river bed, about 12 - 15 m wide and 1.5 m deep was followed closely by the defence line - vallum, to the north of it it was somewhat more distant. Above the level of ground water it was possible to discover relics of wood, which witnesses the fact that some relics of the structure elements of a possible bridge may have been preserved linking up to the eastern gateway (Dostál 1984: 148, Fig. 2).

Geophysical works performed in the space of the northern part of the ringwall on an area of 90 x 50 m and in the network of 1 x 1 m (Voňka 1985: 8-12) were to trace a gateway of a similar type to that on the eastern side. Magnetic measurement (Fig. 74) demonstrated a destructional burned layer of clays of the defence line - vallum body on the whole measured area in the direction SE - N, only locally interrupted approximately at the central part. A position of a smaller gate of a different construction cannot be excluded, such as a stone gate, than in the case of the eastern gateway. This would also be documented by the existence of a road going from the magnate farmstead to this space (Dostál 1970: 12). The existence of a gateway of a similar type as on the eastern side of the ringwall can be quite eliminated in that studied sector.

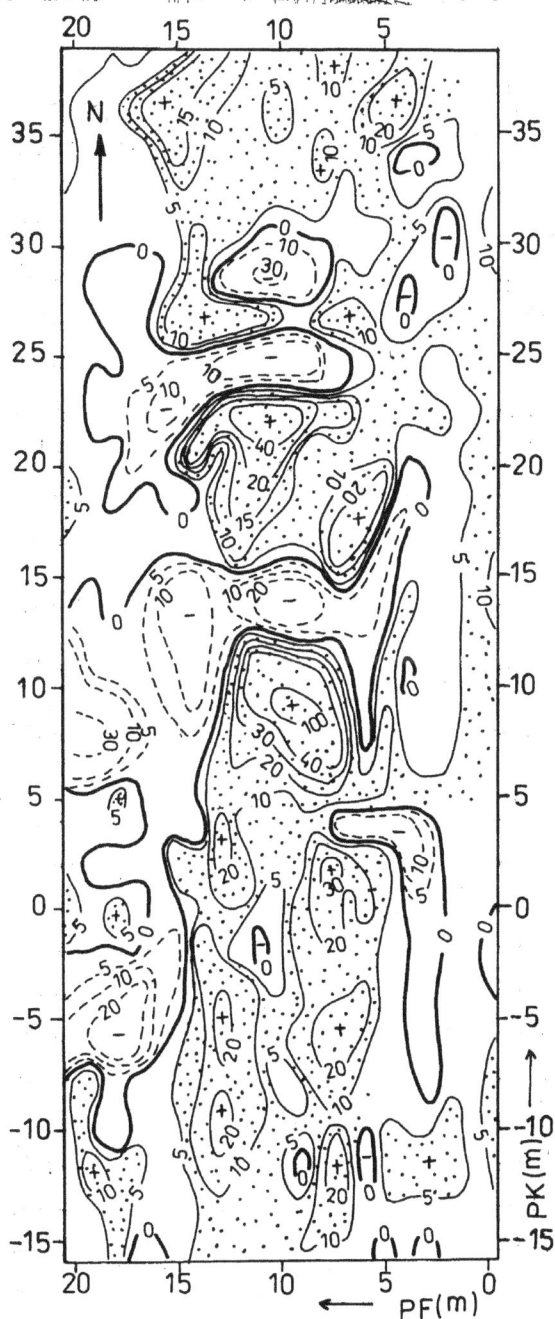

*Fig. 73. a- experimental reconstruction of gate (drawing A. Šik)*

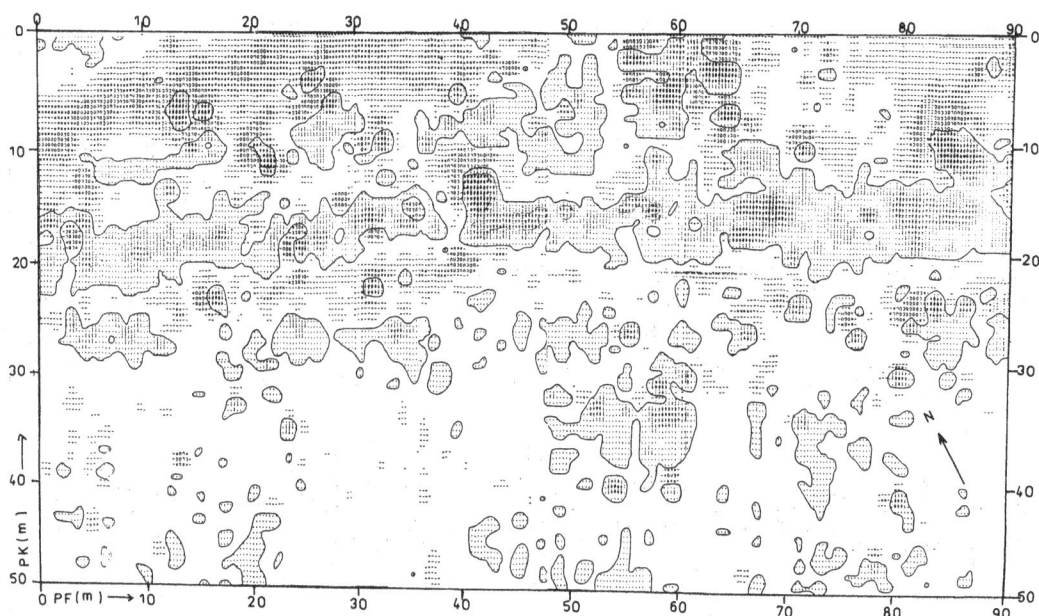

**Fig. 74.** *A shadow map of ΔT anomalies in the eastern part of the stronghold (Voňka 1985).*

In the area of interest, within the ringwall as well as outside it, some extensive anomalies of the magnetic field (+ 15 nT) can be separated which are probably connected with the settlement. According to the situation from the excavation east of the measured area it can be e.g. pit-houses with a wicker-work of walls, wooden stakes and furnaces under which there was a layer of burned loam. Those objects burned with greatest probability in the fire of the defence line - vallum and the burned layers of daub are the cause of those magnetic anomalies. But other sources (storage pits, etc.) cannot be excluded either.

## 4.2.4. Medieval Fortified Settlements

Systematic and rescue archaeological research into medieval fortified settlements (castle, so-called small castle, stronghold, residential farmstead) is concentrated above all on the identification of the ground plan situation of fortification parts, i.e. moats, walls, towers etc. distinct from the habitation element which used to be a palace or a habitation tower. On the basis of results of probing it is possible to eventually draw up a hypothetical reconstruction of the original form of the studied object. Besides this role the investigation also solves further issues, such as the position, size and the ground plan layout of the individual buildings, production and farming objects in the external settlement, wells or tanks, etc.

Geophysical works are carried out on the basis of the results of measurement of physical properties of rocks both by magnetometry and by geoelectric methods (RP, DEMP). They are concentrated on

    a) mapping of the foundation masonry of the fortified area   and its building disposition and/or earlier buildings,

    b) location and determining the character of the object studied, such as a stronghold, fortification elements surrounding some of the above seats, farmstead, relics of earlier built-up area, etc.

    c) finding different stages of building and annexes of the main object,

    d) following the positions of further buildings at the structure studied.

### 4.2.4.1. Strongholds, Small Castles

*Brumovice, EMV Divice, district Břeclav*

Probing works carried out by the Regional Museum at Mikulov in the broad area of the deserted medieval village of Divice (the 13th to the early 15th centuries) at an elevated place called "Kostelík" (i.e. Little Church) demonstrated the existence of a moat with a triangular section along its circumference and on a hill remains of a heavily damaged seat of the "motte" type which originally bore on its upper platform a loam-wooden built-up area (Unger 1988: 210-215). That investigation was continued in 1987 by geophysical works by the magnetometric method (Hašek et al. 1988b) which, after experimental measurement in the vicinity of the verifying probe (Fig. 75) found on the measured area of about 153 x 90 m a moat about 7 to 8 m wide (see Fig. 76) separating an approximately square shaped area of the stronghold consisting largely of a burned layer of daub from the surrounding settlement of the space (Hašek-Měřínský 1989: 103-151).

To the south of the above structure (Fig. 77) in the direction uphill, the prospection discovered a linear formation - probably a moat (?) extending from the the fortification of the stronghold and again joining in an arch the circumferential moat of the studied area.

The entrance to the farmstead, geophysically interpreted between the fortification of the stronghold and the above linear structure (Fig. 78) is supposed at its southernmost part, near the studied relics of the church foundations. Inside the relatively large object several approximately isometric and linearly oriented anomalies ΔT were separated which may be

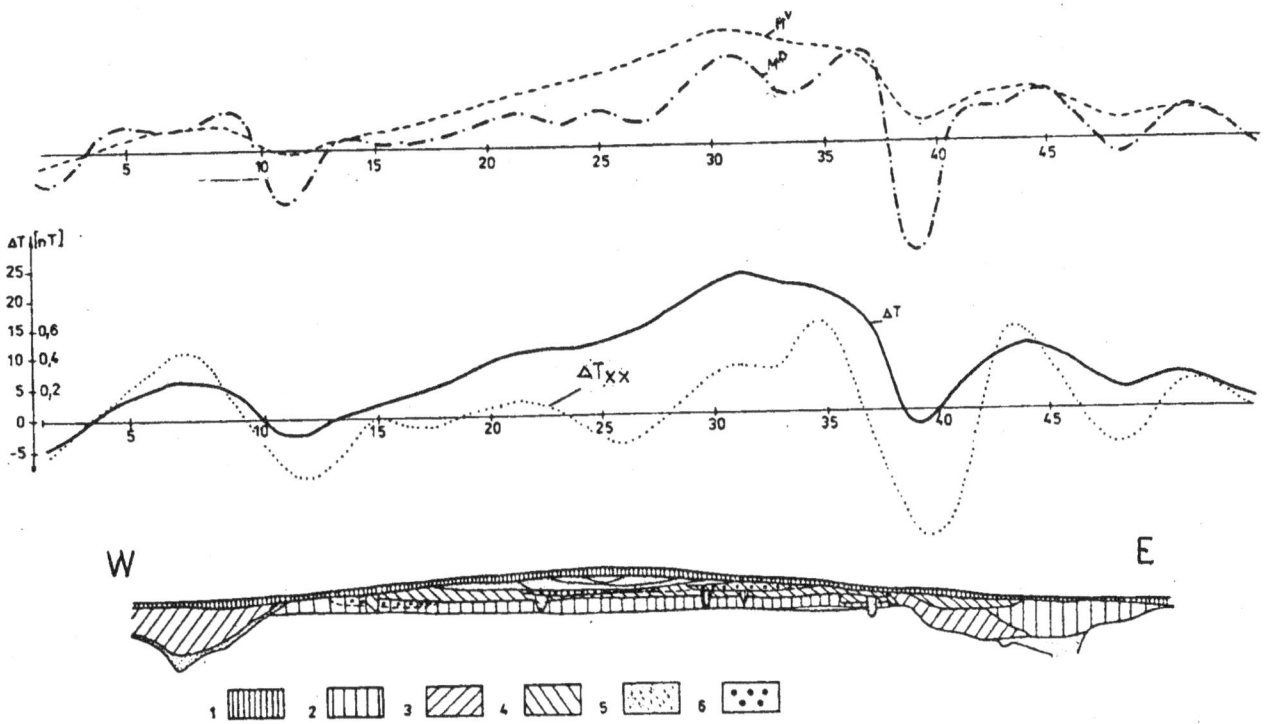

**Fig. 75.** *Brumovice, extinct medieval village of Divice, district Břeclav. A comparison of geophysical results with the archaeological situation. 1- mould, 2- black soil, 3- gray/black soil, 4- gray soil, 5- yellow soil, 6- charcoal, clay (Unger 1988)*

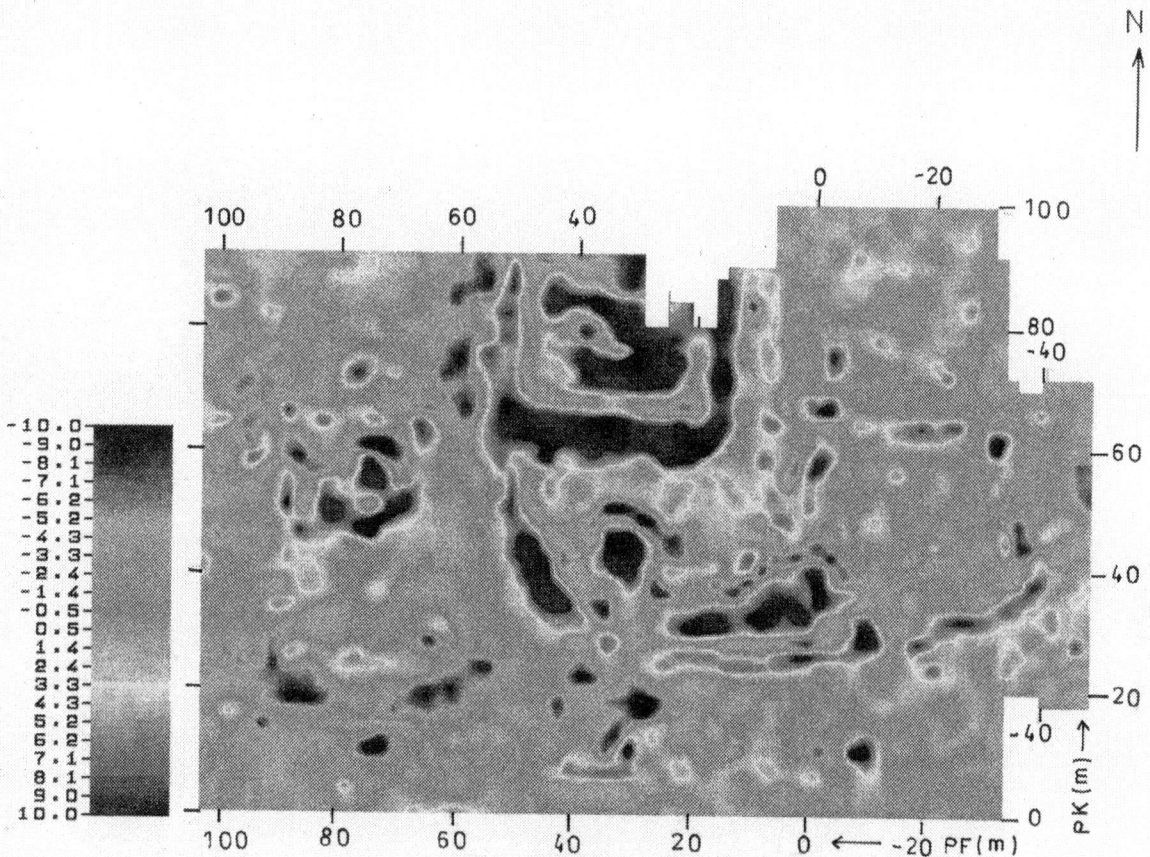

**Fig. 76.** *Brumovice, extinct medieval village of Divice, district Břeclav. Map of ΔT isanomales in the area of the stronghold.*

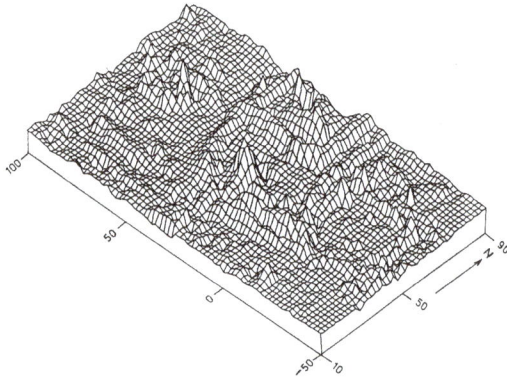

*Fig. 77. Brumovice, extinct medieval village of Divice, district Břeclav. Axonometric depiction of ΔT anomalies.*

connected with its built-up area, i.e. with objects of farming character and habitation buildings.

An extensive occurrence of isometric anomalies ΔT at the western side of the stronghold is worth mentioning. They can be in relation to the extinct village. Besides, there is a course of a narrow linear anomaly included in a circle at the southern and the eastern sides of the measured area (see Fig. 76), approximately simulating another delimitation of the farmstead, evoked probably by the filling of a shallow moat and/or the enclosure ϑ. The overall archaeological verification of geophysical interpretation will be performed in the following years.

Note: Geophysical works carried out by magnetometry in 1989 (Hašek et al. 1989a) W of the hitherto studied area, near the present brook, discovered on an area of about 50 x 100 m probably the positions of 6 habitation and farming houses of the extinct medieval village, oriented in parallel with the discharge trough, i.e. in the NW - SE direction.

*Daskabát, district Olomouc, - "Kopec"*

The studied position is situated at the tract called "Slaný" situated NW of the community of Daskabát. The objective of the geophysical measurement carried out there was to verify the ground plan situation of a small castle of the 13th to 15th centuries in connection with its detailed topographic mapping and the subsequent archaeological investigation. By the methods DEMP and that of vertical gradients (Fig. 79) on the area of about 100 x 40 m it was possible to map both the position and size and the possible spatial layout of the individual building elements for the purposes of the reconstruction of the assumed shape of the construction (Fig. 80).

It concerns an area of an approximately oval shape, divided by a circumferential moat into two extensive areas, a major one of the small castle proper on the western side and a smaller suburbium in the eastern sector of the station. In the wall sector it is possible to expect relics of two to three buildings characterized by regions of reduced conductivities

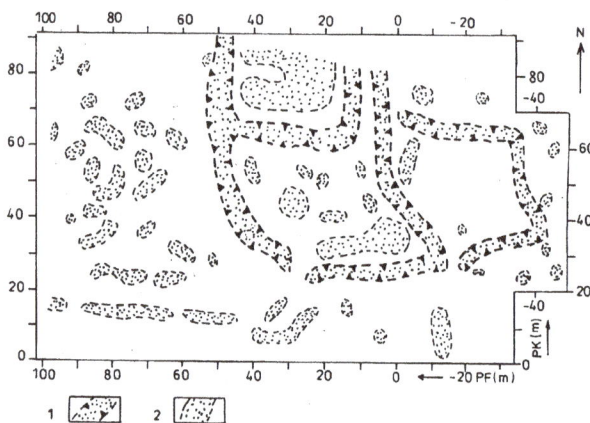

*Fig. 78. 1- ditch alignment, 2- sunken objects*

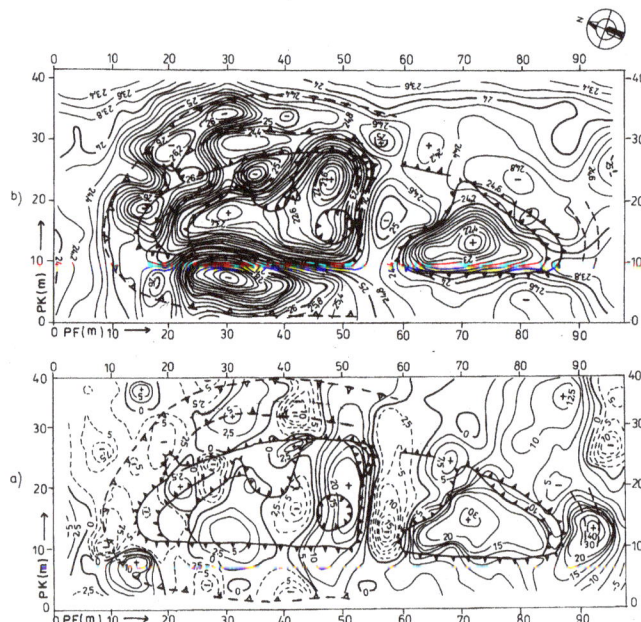

*Fig. 79. Daskabát, district Olomouc, "Kopec". Map of $T_z$ gradient and $\rho_{DEMP}$; interpreted plot of the castle.*

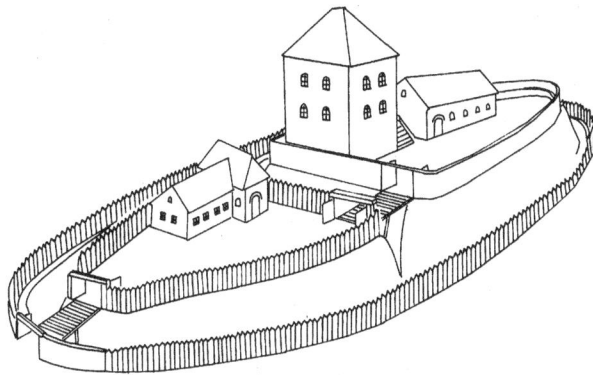

*Fig. 80. Daskabát, district Olomouc. An artist's depiction of the castle.*

and positive anomalies $T_z$. Among the most significant ones it is possible to include the assumed central habitation building (stone structure, daub floor ?) at the eastern margin of the studied area, linked up by another, probably smaller object. At the northern margin of the above area another building may have stood and possibly also a stone wall encompassing the whole are of the size of about 40 x 20 m.

In the smaller sector of the suburbium, dimensions about 25 x 15 m, only one extensive positive anomaly $T_z$ was measured and a region of reduced conductivities (see Fig. 79). As in the fortification proper, this may be a sign of the destruction stone layer maybe also with burned stones and clays from several overground objects.

The two areas are enclosed and approximately in the N- S direction divided by a moat of variable width of 6 to 8 m. But a certain distortion due to later field interventions and adaptations cannot be excluded. At the eastern margin of the suburbium the course of a road into the studied object is interpreted.

*Daskabát, district Olomouc - "Zámčisko"*

Geophysical works carried out by magnetometry and the DEMP method at the wider space of the medieval stronghold "Zámčisko" ("Alt Schloß") situated NE of the community of Daskabát were to verify its ground plan situation in connection with a detailed topographic mapping and subsequent archaeological investigation (Hašek-Bachratý-Tomešek 1994). From the written sources it is not sure whether that station was the seat of Žibřid of Újezd mentioned in written sources to the year 1324 (Hosák 1967; Dohnal 1977) or the seat of the aristocratic family deriving its name from Otěhřib, the predecessor of the present Daskabát in 1281 and 1283. Another thing that is not sure is, whether the report about the deserted stronghold called "Kopec" of 1447 relates to this station (Nekuda- Unger 1981: 231, No. 467).

From the results of geophysical measurements processed into maps of $T_a$ (Fig. 81) and isolines $\rho_{DEMP}$ for h = 3 - 5 m (Fig. 82) in combination with the morphology of the field it was possible to follow the probable ground plan situation of the studied object situated on a small artificial hill.

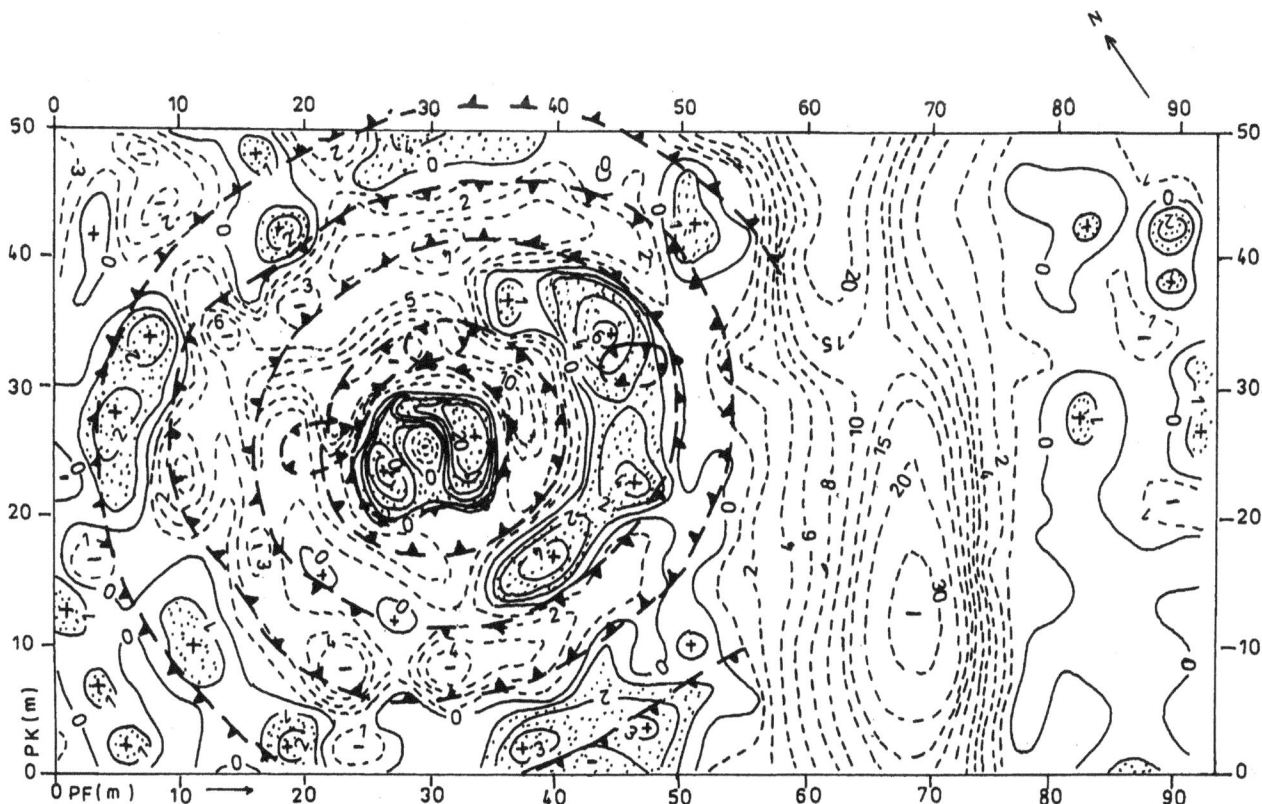

*Fig. 81.Daskabát, district Olomouc „Zámčisko". Map of gradient and scheme of the building.*

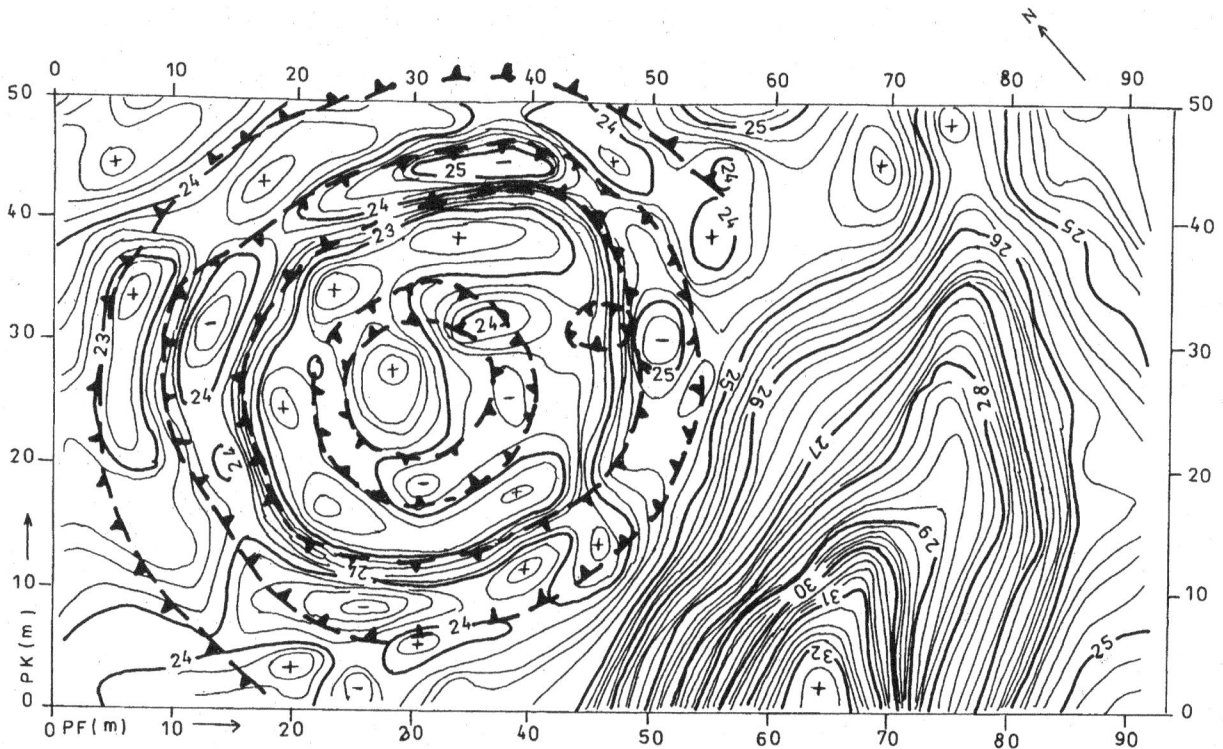

***Fig. 82.*** *Daskabát, district Olomouc, "Zámčisko". Map of* $\rho_{DEMP}$ *isolines and scheme of the building.*

The central, slightly asymmetrically situated, probably tower-like habitation building of the interpreted dimensions of about 8 x 8 m (not even two objects of 8 x 4 m and 3 x 4 m cannot be excluded) may have been built either of wood and destroyed by fire (daub floor) or of stone (destruction layer ?) or of the combination of the two building materials, which is partly indicated by data of the two methods employed.

Along the circumference of the building it is possible to separate a small and shallow moat (diameter about 14 m, width 3 - 3.5 m) which, however, is little conspicuous in the field relief. The outer moat, relatively sharply recessed and conspicuous in the field, has the diameter of 28 m and the width of 4 - 6 m. Its greater recession and widening can be assumed at the western margin of the measured area (see Figs. 81, 82). The bodies of the inner and the outer defence line vallums are marked almost along the whole circumference by characteristic positive linear anomalies $T_z$ and reduced values of $\rho_{DEMP}$. In the inner defence line - vallum besides the earth bank the influence of the daub floor cannot be excluded.

### 4.2.4.2. Castles

*Lelekovice, district Brno-country*

In connection with the overall specification of the ground plan situation of the fortified medieval settlement, on an area of about 840 m$^2$ by physical measurement (Hašek-Tomešek-Unger 1996) the course of the eastern part of the circumferential moat of a 14th century castle was studied (Unger 1990) and positions of further inhomogeneities being in connection with that object were followed.

By the GPR and DEMP methods several anomalous elements were located in the area studied (Fig. 83) that can be assigned to the expression of the circumferential castle moat and further building elements, such as relics of the foundation masonry from the fortification walls, etc. A relatively marked interface of electromagnetic waves was found at times of 12 to 38 ns. The relief of rocks of the Brno Massif (weathered granodiorites) at the selected effective rate of 0.15 m/ns can be found there in the depths of about 0.9 to 2.8 m. The main separated and turning linearly oriented zone of reduced

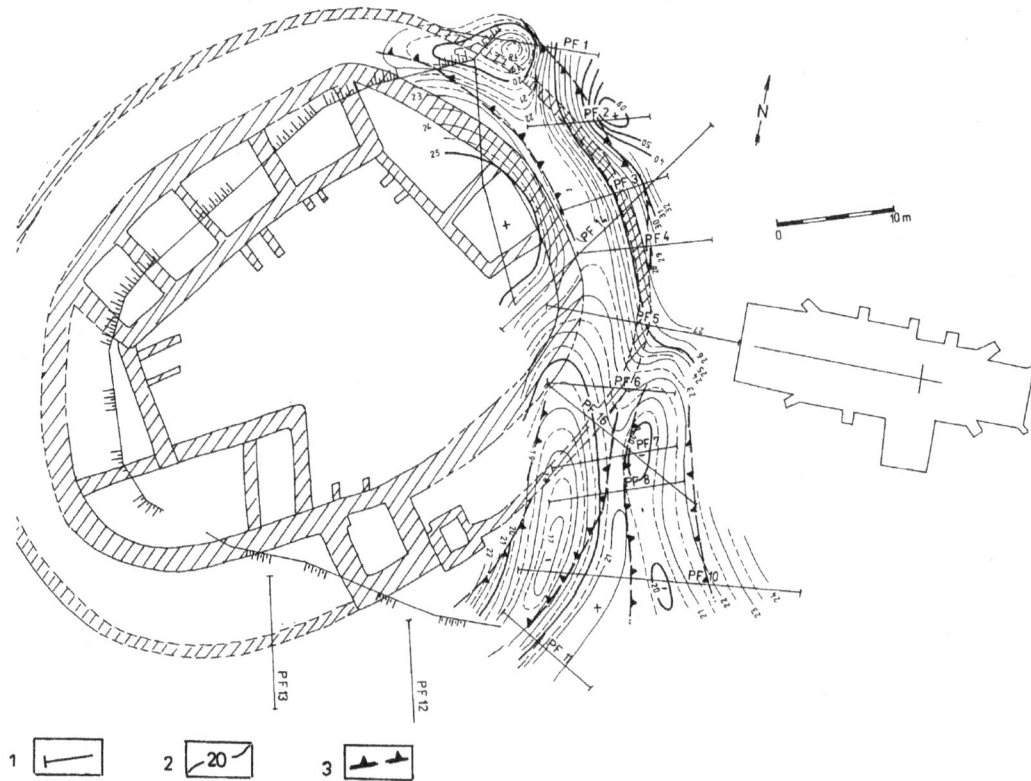

***Fig. 83.*** *1- profiles of GPR and DEMP, 2- isolines $\varsigma_{app}$, 3- outline of the moat*

conductivities is about 6 - 7 m wide, with maximum depth of 2.8 m (see Fig. 83). Despite a lower physical expression the position of the castle moat is interpreted there which was probably recessed prevailingly into the eluvium of the Brno Massif. Its secondary filling consists of stone-sandy material from the extinct castle. A smaller and shallower moat about 2 m wide is assumed in the broad surroundings of the present church of St. Philip and St. James. Further narrower nonconductive zones - relics of the fortification wall masonry can in places be separated particularly at the inner side of the moat.

*Rokštejn, district Jihlava*

The castle of Rokštejn near Panská Lhota (district Jihlava), studied since 1981 by the rescue investigation of the AI ASCS in Brno, and since 1994 by the Faculty of Arts, Masaryk University, Brno, was founded in the course of the 13th century and became extinct most probably violently during the Hussite Wars (Měřínský - Plaček 1989: 4-16; Měřínský 1991: 413-415). Before the beginning of the excavations in 1981 it was necessary to verify geophysically the run of the foundation walls in places where overground masonry had not been preserved and/or follow also remains of objects from the earlier phase of the castle existence (Hašek - Měřínský - Unger-Vignatiová 1983: 146-148, 150; Hašek - Měřínský 1991: 164-166).

In the 1st stage (1981 - 1982) the accessible space of the lower palace was measured by magnetometry and geoelectricity (SRP A2M1N2B m) in the network of 1 x 1 m as well as the court of the lower castle and the assumed fortification line in the southern and the southeastern part of the lower castle (Hašek - Měřínský - Vignatiová 1982: 15-16). From the results of geophysical works it followed that the foundation masonry is mostly conspicuously reflected both in geoelectricity (narrow zones of increased specific resistances) and in magnetometry (æ = 1.3 - 1.6 . $10^{-3}$ u.SI). A very conspicuous coincidence between the results of geophysics and the excavations was found in the main fortification wall in its southern and southeastern course which can be correlated from the lower palace up to the tower on the eastern side (Fig. 84). Relatively intense anomalies of the magnetic field near the above eastern tower of the lower castle are connected with destructions and positions of foundation walls of further objects (Hašek-Měřínský 1987).

The second stage of geophysical works in 1984-86 was concentrated on the space of the upper castle. The separated anomalies ΔT (+ 10 nT, + 15 nT) in the eastern part are signs of the main fortification wall which also formed the base of the eastern wall of the palace of the upper castle, and in the south it started from the tower of the gateway of the upper castle. In parallel with it, but more to the west there is another foundation belonging to an earlier encompassing wall fortifying the castle on the eastern side, dating back to the last third or quarter of the 13th century. The measurement inside the palace of the upper castle confirmed the assumed course of the foundation masonry of the northern wall of the

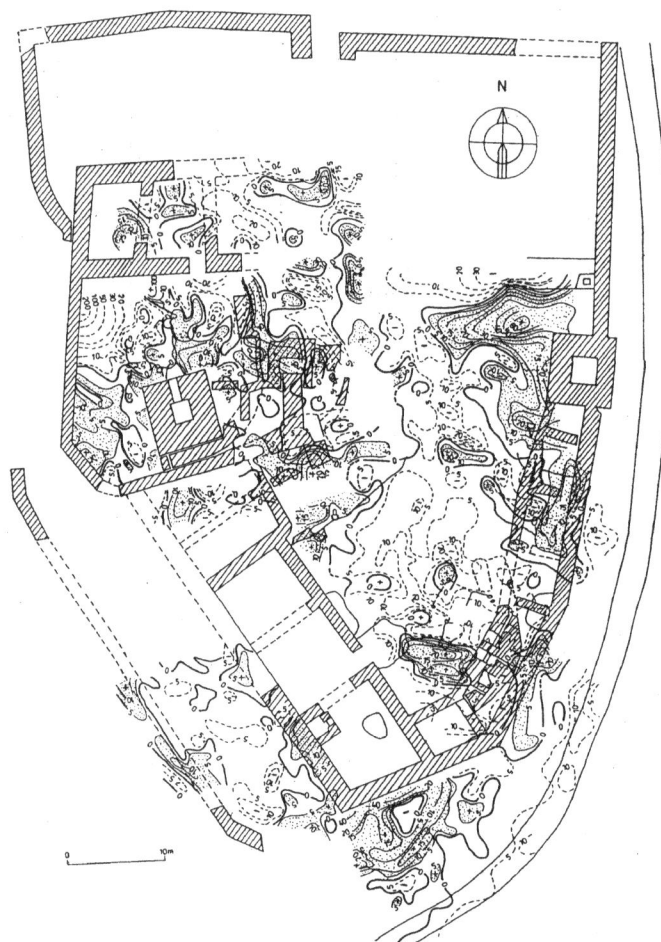

*Fig. 84. Rokštejn Castle, Panská Lhota, district Jihlava. Map of ΔT isanomales; general situation of the castle.*

palace. Conspicuous also are two approximately isometric anomalies ΔT (+ 10 nT) in the court west and north of the tower of the upper castle which are evoked, as found by the investigation, on the one hand by an extensive recessed basement space of the original palace construction of the castle of the last third or quarter of the 13th century to the first half of the 14th century west of the tower, further in the space of the castle court north of the smaller tower, a pit house dug in the rock and a water tank. The course of the moat around the core of the 13th century castle is no longer so marked from the results of magnetometry. From the character of the anomaly ΔT (+ 5 nT) it is possible to follow it in its whole southern and eastern course (see Fig. 84, cf. Měřínský - Plaček 1989: 17-24; Měřínský 1991: 415-419).

*Valtice, district Břeclav*

The task of geophysical prospection was to verify the extent and the possible dispositions of the Romanesque-Gothic structure which had stood in front of the park face until the beginning of the 18th century and was pulled down on the completion of the culminating Baroque reconstruction (Hašek - Měřínský - Plaček 1996: 119-124). To solve the upper set of issues, the DEMP method (h = 1.5 m) was employed on the area of 7616 m² (136 x 56 m). Its results were utilized in the subsequent castelogical interpretation (Fig. 85).

The separated zone of reduced resistances in the form of an irregular pentagon, about 12 m wide (see Fig. 85) evidently represents an artificial depression - a moat delimiting the area of the castle proper. From the middle of the western side another region of reduced resistances starts into its space, turning towards the east, 8 - 10 m wide, in which it is possible to express the assumption of the position of a shallow moat. The different character of the separated places with increased specific resistances may possibley hint at a varied character of the built-up area or the stages of construction. In the southern sector an extensive area of increased resistances was found witnessing a thick stone destruction from the complex of buildings at the fortification wall (Hašek - Měřínský - Plaček 1996: 125). At the northeastern part of the studied area several zones of increased resistances were discovered, evidently due to the relics of masonry, among them two linear belts of reduced resistances above depressions. In the south, behind the moat there is a longitudinal formation of increased resistances (the solidified position of the mount, masonry destruction), from its eastern end at an acute angle a linear zone of lower resistances starts which, like the areal increase in the southwest, starts from the measured area. The situation also indicates the possibility of existence of a minor suburbium.

From the overall course of the values of apparent specific resistances and the iconography it is possible to interpret the

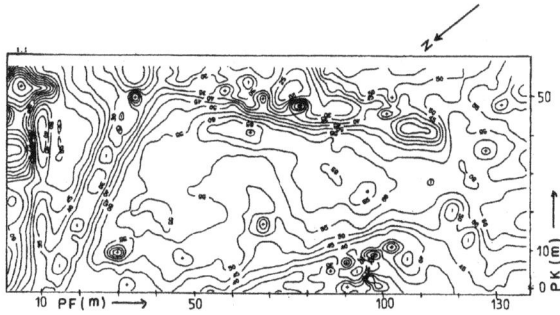

*Fig. 85. Valtice, district Břeclav. Map of $\rho_{DEMP}$ isolines.*

### 4.2.4.3. Military Camps and Modern Forts

*Božice, district Znojmo*

positions of further buildings of the circumferential built-up area, even though they grew most probably gradually and the possible location of the entrance to the castle below the tower from the southwest and northeast. Since there exist hints of the division of the castle by a moat across it from the west to the east and the central object which would otherwise be an obstacle to the entrance from the northeast, at the station studied the assumption of the archaeologists seems to be confirmed that this stone, fullest medieval castle had a probable predecessor in a two-part feudal seat of maybe wood-loam design with a habitation tower (Hašek - Měřínský - Plaček 1996: 126-127. Between the moat and the park face of the chateau (see Fig. 86) the character of specific resistances hinted the position of the destroyed part of the Renaissance chateau which corresponds by its width with its structures found by the investigation in the basement of the chateau building.

From the data of aerial prospection in 1983 (Kovárník 1993: 108-110) a double circular formation was discovered on a mild hill near the community. For the overall specification of its position, size, etc. on an area of 80 x 90 m magnetometric measurement was done which, according to positive linearly oriented anomalies $T_z$ indicated the possible existence on the one hand of an inner, wide circular moat, diameter about 38 m, interrupted by probably only one entrance in its northeastern sector, and, on the other hand, an oval narrower moat with the long semiaxis of about 75 m, oriented approximately in the N - S direction (Fig. 87). Inside as well as outside the separated structure were several positive isometric anomalies of a magnetic field turning into a circle, which were interpreted as probable recessed objects of different character, such as large grave pits or, in the case of a local circular body at the southwestern margin of the measured area, the possible position of a barrow, etc.

Archaeological investigation confirmed the existence of a double circular area. It proved that the two circles were built in the first third of the 15th century. Besides pottery material of ancient age (Eneolithic, the Bronze Age) the inner moat contained an indispensable part (fragments of vessel margins) of the pottery of the 1st third of the 15th century. They are the youngest finds dating the moats.

Circular moats can be interpreted as light field fortifications built probably by Hussite troops or by their enemies in their campaign in south Moravia (Hašek-Kovárník 1996: 73-75).

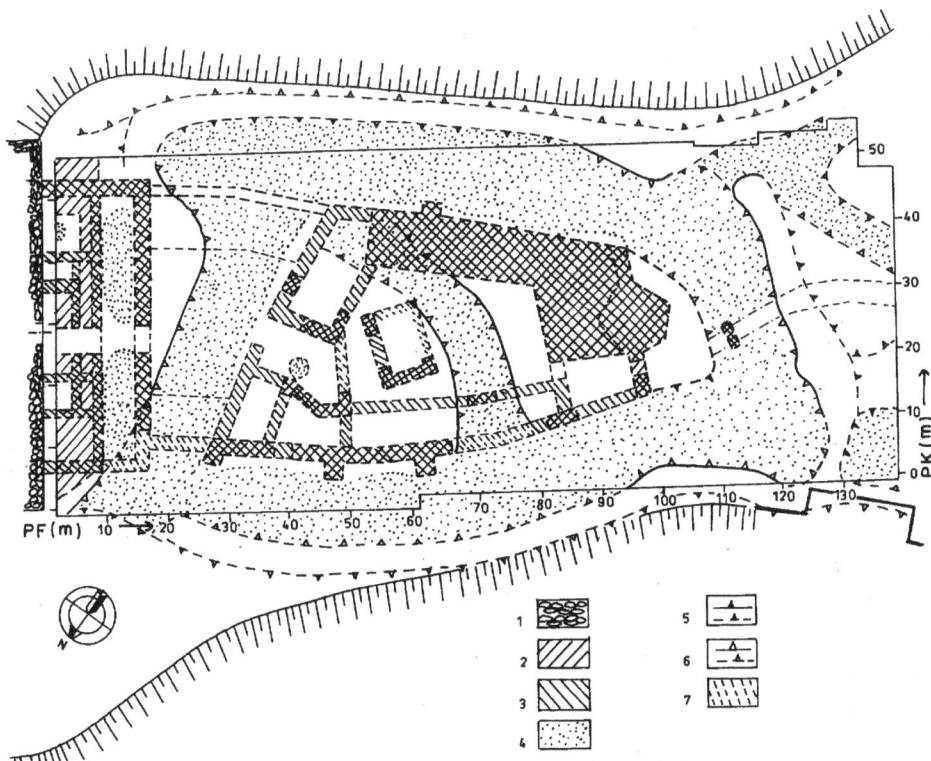

*Fig. 86. 1- chateau, 2- masonry assumed by geophysical measurement, 3- completion of the masonry with respect of the results of geophysical work, 4- depression, 5- field edges according to the geophysical data, 6- field edges completed, 7- completion of masonry (interpolation and iconography)*

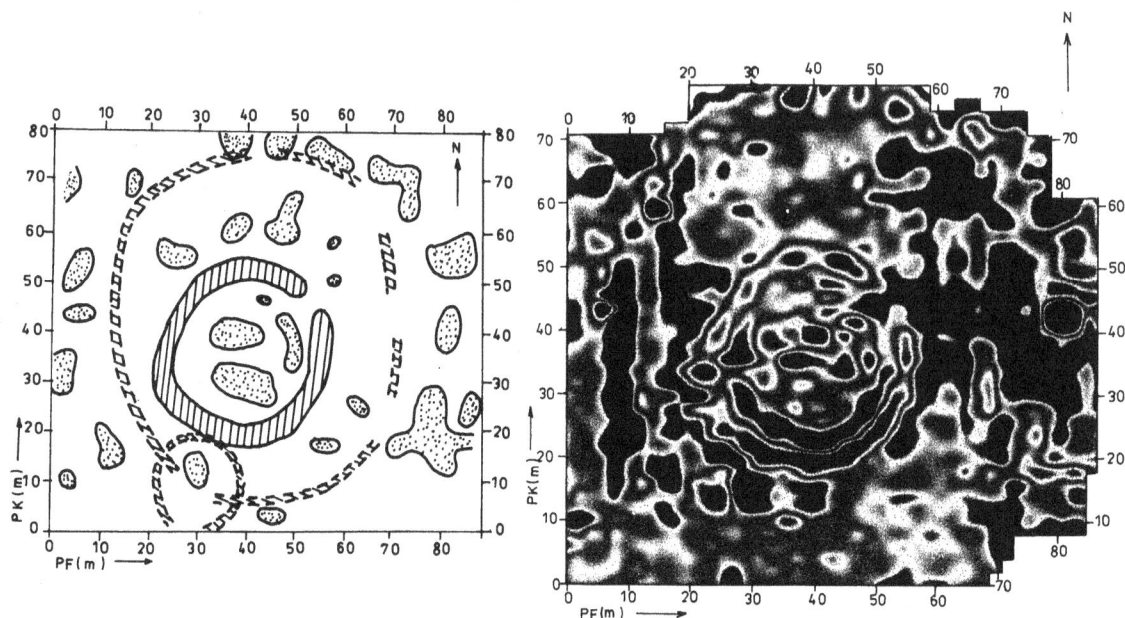

**Fig. 87.** *Božice, district Znojmo. Surface layout of $T_z$ gradient and correlation scheme of geophysical results.*

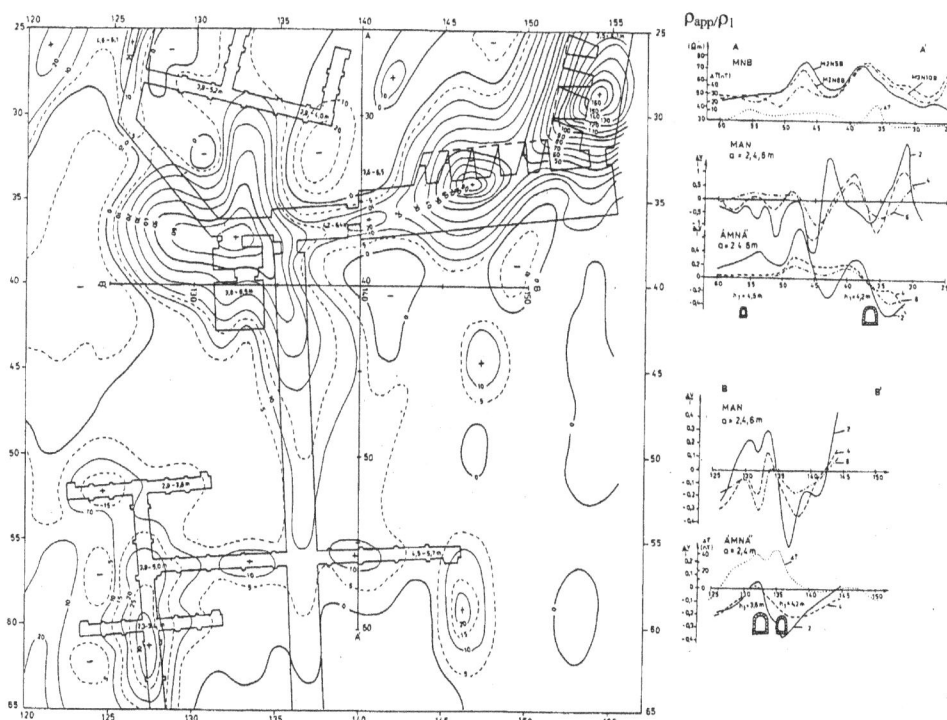

**Fig. 88.** *Jaroměř-Josefov, district Náchod. Map of $\Delta T$ isanomales, comparative geoelectrical profiles A-A' and B-B' over corridors.*

The outer moat of mildly triangular section is 1 - 1.2 m wide, 0.80 m deep, diameter 72 - 76 m, the inner (main) moat has a dish-shaped profile of the width of 5 - 7.3 m, depth of 0.8 - 1.1 m, (inner and outer) diameters of 27 and 39 m, respectively. The study of the knoll brings also objects of an earlier period. In the course of the investigation a cremation grave of the Bell-Shaped Goblet Culture (2500 - 2000 years B.C.) was investigated, two inhumation graves of the older period of the Únětice Culture (1900 - 1550 years B.C.), a grave of the Mid-Danubian Barrow Culture (1550 - 1300 year B.C.) and 16 Slavonic graves of the Late Ringwall Period (11th century). Approximately in the western sector inside the moat a system of foundation grooves was found. From them a ground plan of a building of rectangular shape can be reconstructed with dimensions of 3.5 x 7.5 m. It is possible that another building of that type may have existed there (Hašek - Kovárník 1996).

Jaroměř - Josefov, district Náchod

An important member of the Theresian fortress Josefov (1780-1787) is the system of extensive underground corridors which were formed in cretaceous rocks in three storeys. The underground spaces are vaulted and lined with burned brick. They stretched on the one hand under the main bank, on the other hand under the glacis. Under the main defence line - vallum the corridors were interconnected into large spaces, under the glacis they led either to the mine chamber in the glacis or to the listening chambers. The depth of the vaults below the surface varies from 2.5 to 5.5 m. Their dimensions are from 0.8 to 2.5 m, the height of manhole corridors with the clearance of 1 m and chambers only 0.52 m high up to meeting-places 3.1 to 4.5 m high.

The objective of geophysical works was to verify the possibility of using magnetometry and different modifications of resistance profiling for solving the issues of the above type of cavities under conditions of urban agglomerations (Hašek - Odstrčil - Pantl 1980). Magnetic measurements carried out in the network of 1 x 1 m located the course of the lined corridors of different types relatively reliably (Fig. 88) and from geoelectric profile measurements (AMNB, AMN -B, MAN and A MNA -B) with the step of 1 m it followed that symmetric resistance profiling was not too effective for following that type of cavity in the above depths, unlike the differential arrangements (MAN and MNA -B), by which those inhomogeneities were found relatively well (reflection points of curves V/I) even in depth of 4 - 5 m (see Fig. 88).

## 4.3. Investigation of Ritual and Sacral Buildings

The investigation of prehistoric ritual areas in Egypt and medieval sacral buildings by the methods of geophysical prospection is directed above all on finding the course of the foundation masonry, its ground plan layout and/or the recession of the individual objects studied as a consequence of their possible static provision from the possible impairment.

The issues of ritual areas were followed by geophysics under conditions of arid climate on the area of the Memphis necropolis near Abussir (Egypt). They concern the solution of some issues of existence of architectonic elements of royal tombs of the pyramid type. These tombs consisted of several big structures and not only of the limestone pyramid. The whole pyramid complex began already at the margin of the Nile valley by the so-called valley temple with a port. Its second component was a rising stone road serving as a ramp for overcoming the height difference between the Nile valley and the plateau. On top of the rock plateau the rising road led to the mortuary temple, the actual place of the mortuary cult. By its western side the temple adjoined the eastern side of the pyramid. The building material for the above buildings were above all blocks and slabs of limestone, granite, basalt, gabbroamphibolite, sandstone, further dried brick of the Nile mud, etc. (Hašek - Obr - Přichystal - Verner 1986: 149-187; Hašek - Obr - Verner 1988: 5-37). These rocks differ relatively a great deal by their resistance properties and their magnetic susceptibility from the surrounding environment - the Sahara sand which is the covering element for the objects studied. For instructivity it is possible to state that e.g. the

mean value of the Sahara sand - $0.21.10^{-3}$ SI, dried brick - $1.57.10^{-3}$ SI, granodiorite $9.61.10^{-3}$ SI, basalt - $10.4.10^{-3}$ SI, etc. On the basis of the results from submitted laboratory measurements it is possible to use magnetometry completed by geoelectric methods, particularly RP and DEMP in the sectors of interest as the basic method for solving the individual tasks of those issues.

### 4.3.1. Prehistoric Mortuary Temples

*Abussir - Egypt*

On the basis of previous archaeological investigations at the Unfinished Pyramid (sector C, see Fig. 116) (Borchardt 1910; Maragioglio-Rinaldi 1970: 176-184) a hypothesis was expressed that in front of its eastern side there might be relics of the mortuary temple and from it also relics of a rising road towards the east. At the same time Maragioglio-Rinaldi (1970) expressed a guess that due to a large number of rough cobbles on the upper surface of the unfinished core of the object the works at the pyramid were interrupted and in order to "clear up" the area the unfinished structure was covered with sand. These issues were also connected with several hitherto unanswered questions, starting with the name and fate of the owner and ending with the existence or non-existence of the temple. On the basis of the above probing Borchardt gathered (Borchardt 1910) that the pyramid had not a finished sarcophagus chamber, on the other hand it was ascribed to a ruler whose mortuary cult is documented in contemporary written sources (Mariette 1889: 224-295) and it must thus have been held in that mortuary temple. At the same time the existence of such cult would assume its finishing at least to such extent as to make it operative from the ritual point of view (Hašek - Obr - Verner 1988: 5-37).

Serious historical issues connected with this Unfinished Pyramid, conclusions of older investigations and observations, and particularly the surface investigation of the space of the hypothetic temple on the east side of the pyramid resulted in the fact that this very place was chosen in 1978 for a detailed geophysical measurement (Hašek et al. 1979; Hašek - Verner 1981: 306-316). On the area of 60 x 40 m magnetometry was used in the network of 2 x 1 m which was completed by resistance profiling in sectors of interest.

From the resulting processing of geophysical data in the form of a map of isanomales $\Delta T$ (Fig. 89) a relatively broken course of the magnetic field is evident which, already according to positive isanomales $\Delta T$ (+ 30 nT) confirms the fact that in the broader surroundings of the eastern side of the Unfinished Pyramid there is a number of brick structures of different width with axes in the E - W and N - S directions, which in the direction to the east recede to the axis of the pyramid.

According to the correlation dependences of the individual anomalies $\Delta T$, linking up with the probing works and, last but not least, the areal excavation, a relatively extensive and on the whole little damaged object (see Fig. 89) was located in the measured area. It was built prevailingly of dried brick ($\text{æ} = 1.1 - 2.0 \times 10^{-3}$ u.SI). Its depth below the present sand surface is about 0.3 - 0.5 m. The exposed masonry reaches the height of more than 2 m in some places.

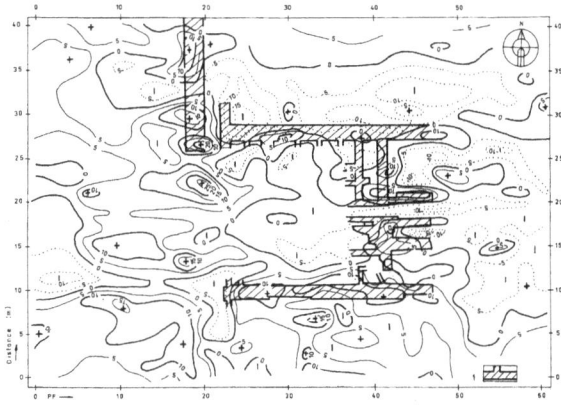

*Fig. 89. 1- building relics*

The building arose gradually in several stages. The circumference of the western oldest part of the temple (Fig. 90) is delimited by a double brick wall more than 2 x 2 m thick, the eastern part is delimited by a wall up to 2 m thick ($\mathrm{æ} = 2.0 \times 10^{-3}$ u.SI). The inner space of the object is divided by walls of different thicknesses into smaller spaces up to individual rooms determined for different functions (habitation room of the priests, stores, etc.). Besides the above structures also preserved door vaults were found there, relics of a staircase, etc. (Hašek-Preuss 1987: 153-158).

Rich written finds including the discovered archives of papyri demonstrated that it is really the looked for mortuary temple of King Ranephereph of the Vth dynasty of the Old Empire (Hašek - Obr - Přichystal - Verner 1986: 149-187).

By the overall confrontation of the results of geophysical works with archaeological investigation it followed that the magnetometric method was able, from the course of positive anomalies $\Delta T$, to indicate not only the existence of the building and its rough outline (Hašek - Verner - Obr 1983: 187-199), but that it was also suitable to give information about its medium rough articles. This, of course, does not hold for all situations in the same way. The reasons of those non-uniformities can be seen, as shown by archaeological investigation, above all in the irregular distribution of the wall destructions of unburnt brick, in later interventions into those destructions and partly also in the different state of preservation of the individual parts of architecture. Negative anomalies of the magnetic field are met above all in places where relics of architecture are situated for whose construction white limestone from the eastern bank of the Nile was used. Another case indicated by negative anomalies is the break through mighty layers of brick destructions by a later intervention. This category includes even those cases when the negative magnetic anomaly locates the places of deposition of secondary burials. And, finally, more extensive but less conspicuous anomalies $\Delta T$ indicate major areas not built up (Hašek - Obr - Verner 1988: 5-47).

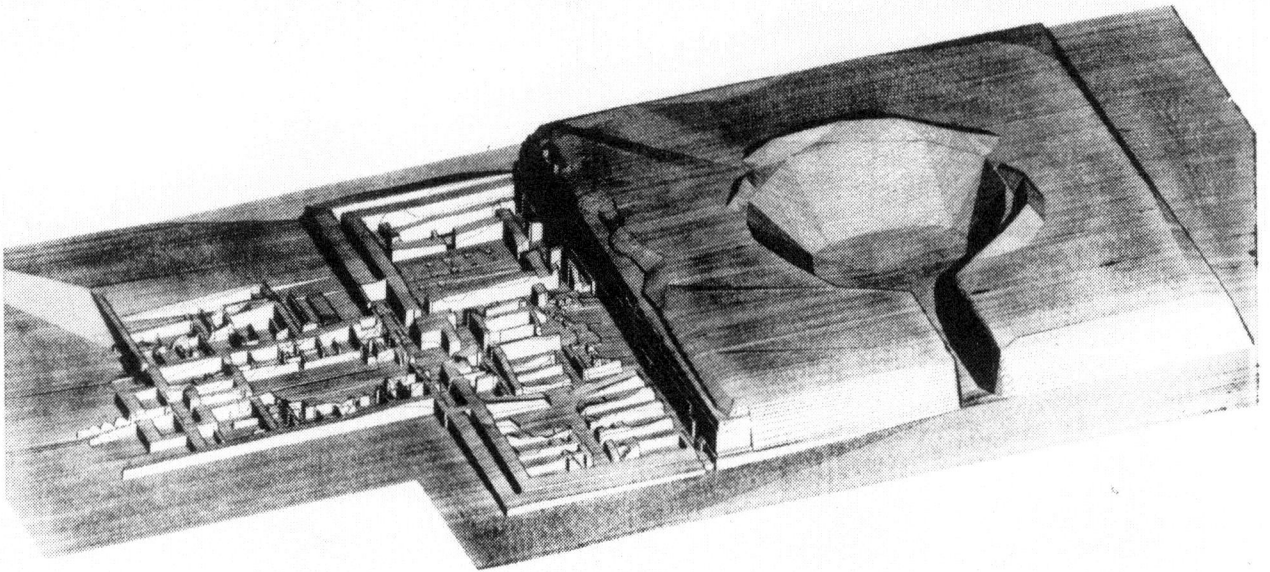

*Fig. 90. Abusir, Egypt. Axonometric depiction of Raneferef's temple (research of the Czech Institute of Egyptology Prague in 1984, Eisler, Pejša, Preuss 1988).*

*Býčí skála near Adamov, district Blansko*

In connection with systematic archaeological investigation of sediments of the cave Býčí skála (Bull's Rock) in the valley of the Křtinský Brook in the Moravian Karst, in the Chamber of the above cave, northeast of its upper or lower entrance and in places of field adaptations dating back to World War II, geophysical radar measurement and DEMP were carried out (Hašek - Přichystal - Tomešek 1996). The main objective of the works was to determine the extent or the thickness of the cultural layer of the ritual place of the Halstatt Period.

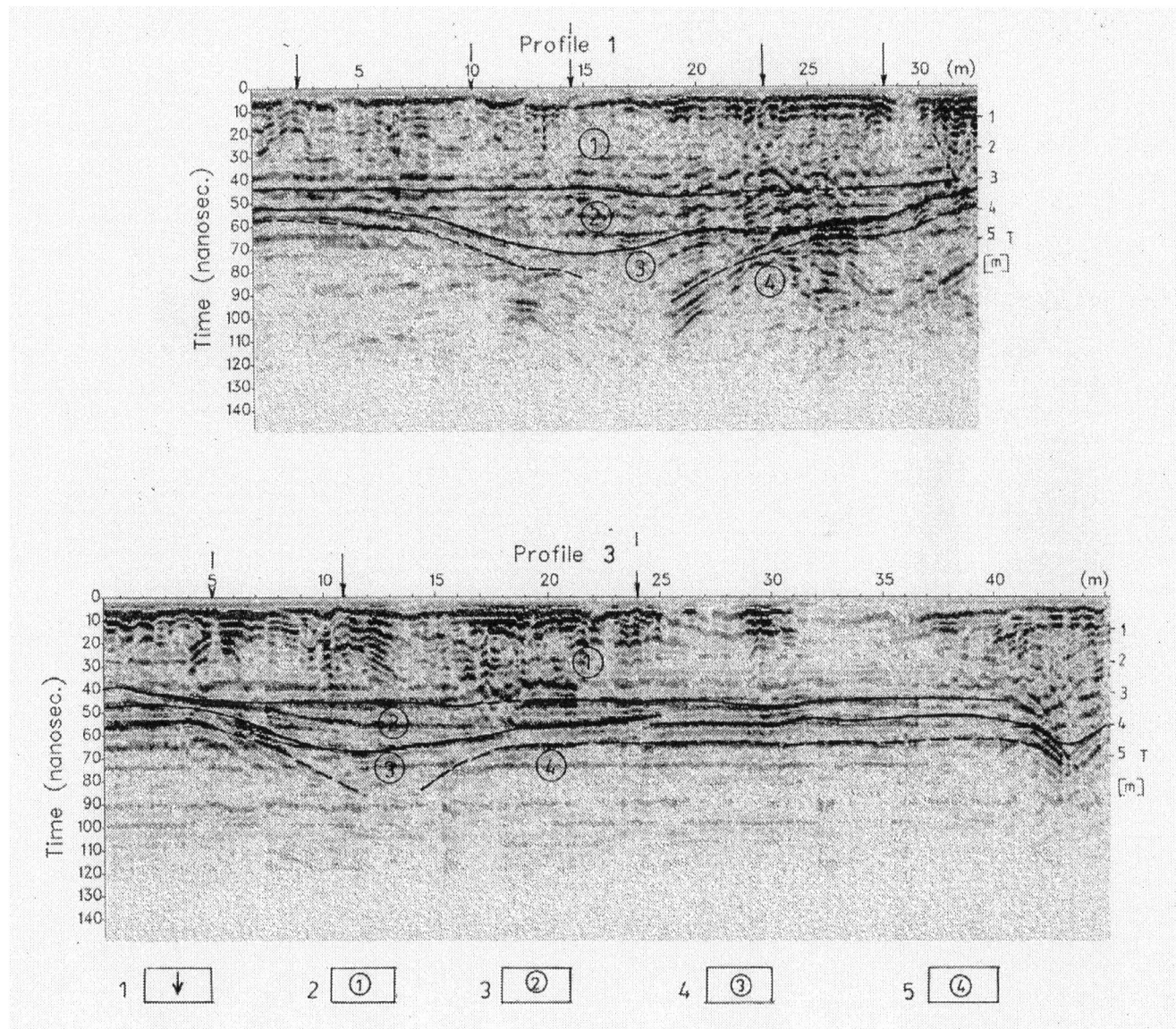

**Fig. 91.** *1- near-surface inhomogenity, 2- concrete floor, made-up ground, 3- sands, layer with signs of cultural development, 4- cave loess (clayey sediments), 5- peleozoic (clastics, limestones)*

The processing of the measured data in the studied space found several marked interfaces of reflected electromagnetic waves (see Fig. 91) taking place at times of 40 - 50 ns, 60 - 78 ns and about 90 ns. In the first case, after comparison with the data of boreholes BS-1,2 it can be a concrete floor and recent made-up ground ($v_r$ = 0.12 m/ns) with maximum thickness of about 2.9 m, in the second case gravel-sands with a possible cultural layer ($v_r$ = 0.10 m/ns) of an approximate thickness up to 1.5 m and in the third case a relief of limestone rocks of the Palaeozoic ($v_r$ = 0.15 m/ns) in the substrate of sandy sediments (cave loess?) at the depth of about 6.8 m. From the made up time map (Fig. 92) and the values of æ, there evidently follows a larger areal extent of the above cultural layer than assumed in the original archaeological evaluation (Přichystal 1993: 75-86). The separated local inhomogeneities are concentrated only into the near-surface layers of the made-up ground (see Fig. 91). They can be assigned to the expression of Fe-wires in the concrete, large blocks of limestones in loams, etc.

The implementation of the archaeological investigation will be more difficult by the existence of a thick layer of made-up ground and in places also a concrete plate in their overlying layers. This recent concern can be expected about the whole area of interest of the studied cave.

The circumferential wall was probably interrupted by the entrance to the area in the middle part of the northern enclosure (see Fig. 94). The overall size of the built-up area is about 600 m².

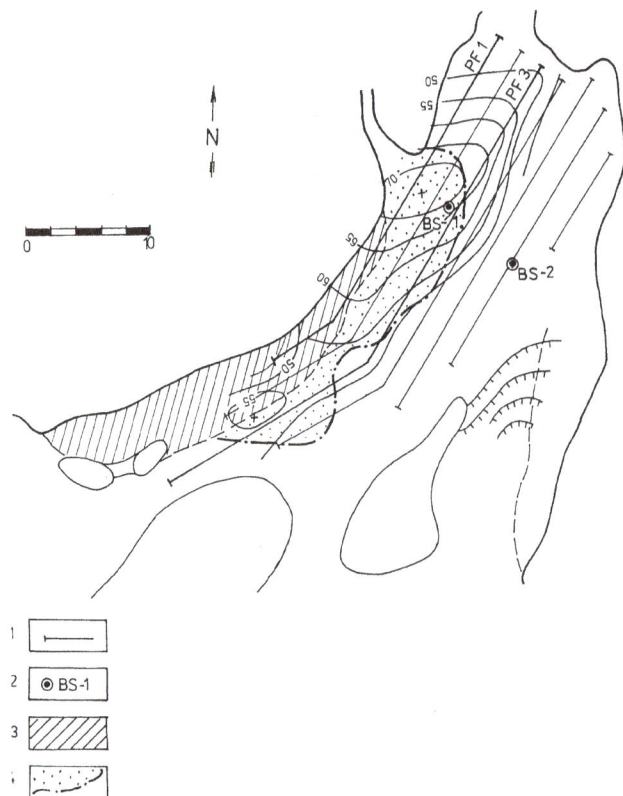

*Fig. 92. 1- profiles with GPR and DEMP measurements, 2- drill holes, 3- uncovered after shooting off roct cover, 4- assumed scope of layer with signs of cultural development*

### 4.3.2 Medieval Sacral Architecture

#### 4.3.2.1. Curches in Towns

*Brno - Královo Pole*

In the city borough of Královo Pole, in the place of the cast iron cross at Mojmírovo Square there used to stand the cemetery chapel of St. Vitus, consacrated in 1279 and falling out of use in 1785.

The purpose of geophysical works by means of the soil radar was to verify the ground plan situation of this building, because neither its position nor its size are exactly known from written sources.

The GPR measurement in a network of perpendicular profiles according to multiple reflections of electromagnetic waves in the form of curves placed below each other, similar to one-arm hyperbolae of different width and orientation (Fig. 93) demonstrated the existence of foundation masonry both from the object itself and probably of the enclosure wall of the assumed width of about 1 - 1.2 m. The depth of the wall relics is about 0.8 to 1.2 m.

The building of the dimensions about 20 x 9 m (Fig. 94), a small apse cannot be excluded, is divided into three partial sections, the choir and a double room of the nave. Later building adaptations can be assumed.

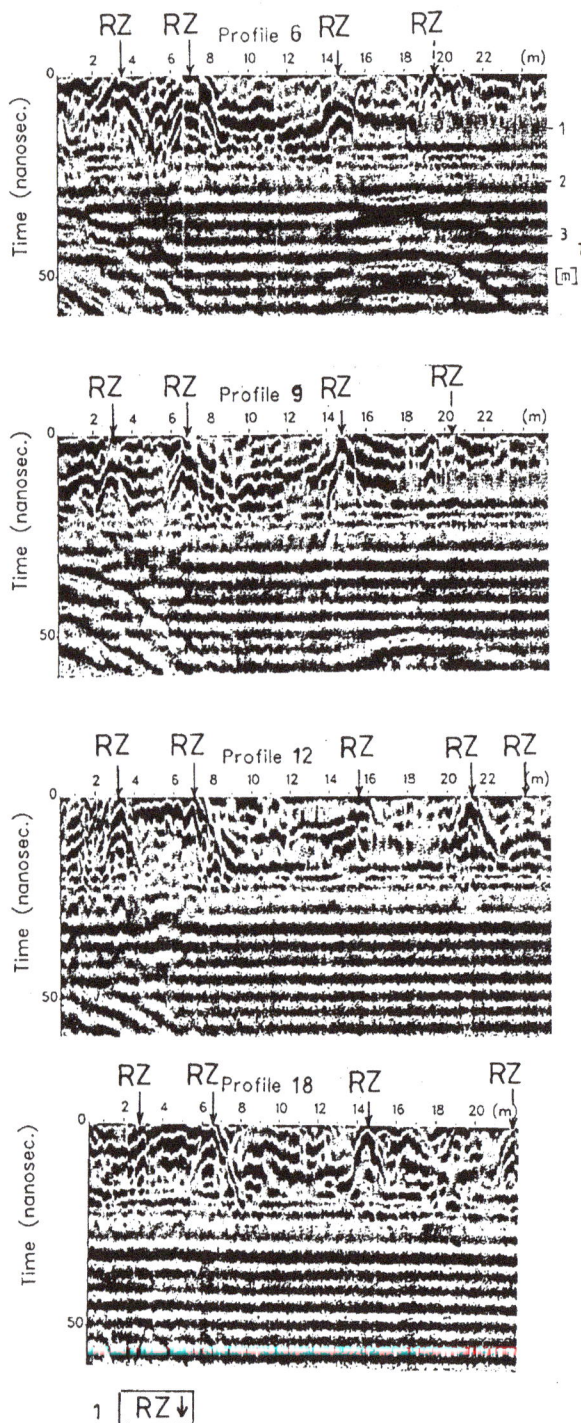

*Fig. 93. 1- interpreted inhomogeneity*

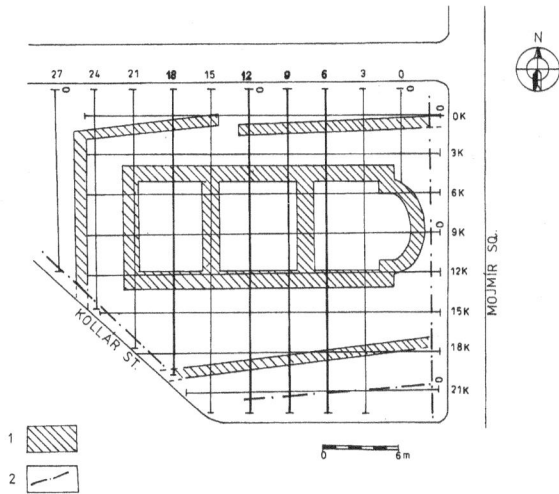

**Fig. 94.** Brno - Královo Pole. Plot of the localized St. Vít chapel.

*Šumice near Uherský Brod, district Uherské Hradiště*

The purpose of geophysical works in the interior of the parish church of Vigin Mary's Birth at Šumice dating back to the early 19th century (Hubatka 1996; Hašek - Pavelčík - Tomešek 1996b) was to find the possible relics of the foundation masonry of an older, probably sacral object and to locate the possible position of a tomb or a grave in the places of the original building.

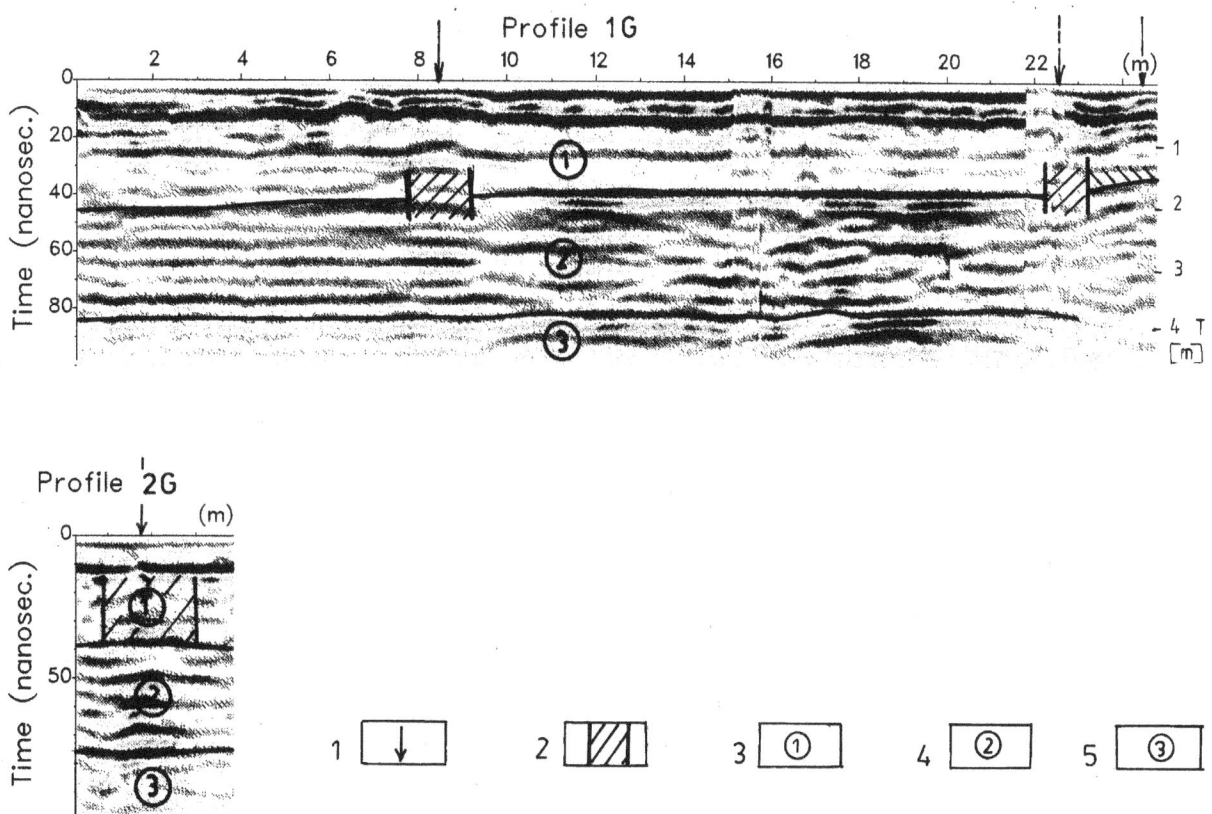

**Fig. 95.** 1- masonry relicts, 2- areal inhomogenity, 3- antropogenic sediments, 4- clayey soils, 5- gravel sands

By the GPR method (Fig. 95) the overall thickness of anthropogenic depositions ($v_r = 0.095$ m/ns) was found into the depth of about 1.9 m and clayey loams to loams ($v_r = 0.125$ m/ns) about 3.4 m in the overlying beds of gravel-sand. The separated local inhomogeneities are concentrated into the depth of prevailingly 0.8 - 1.7 m, i.e. to the layer of anthropogenic deposits. Their interpreted width is about 0.8 - 2 m. The source of shallower "anomalies" can be the foundation masonry from the pulled-down building, in deeper positions it is probably the sign of a grave, tomb, etc.

In the church interior it is possible, according to their layout at the inner side of the circumferential masonry of the existing building (Fig. 96), to interpret the position of an older, probably sacral object of the size of about 5.5 x 11 m. The choir with its dimensions approximately agrees with the present built-up part (about 5 x 3 m). In the space of the nave of the above building no anomalous elements indicating the places of possible graves were found by (GPR, DEMP) measurements. The situation is, however, different in places of the separated chancel, where it is possible to locate a region of reduced conductivities, accompanied by a really more extensive inhomogeneities on the profiles GPR (see Fig. 96). We assumed there was either a small tomb with foundation masonry from the original or other older building, or a grave covered with a large stone slab. Archaeological investigation confirmed our interpretation conclusions. In the above places a grave with a secondarily deposited stone slab was found by excavation in the depth of about 1.4 to 1.8 m.

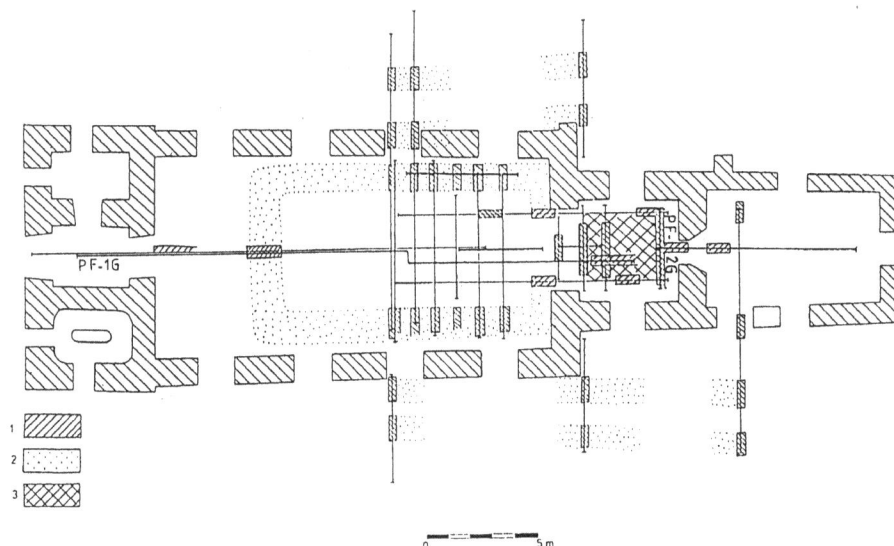

**Fig. 96.** *1- local inhomogenities (fondation masonry), 2- interpreted position of older building, 3- areal inhomogenity of lower conductivity (stony slab, grave, tomb, etc.)*

*Velký Újezd near Moravské Budějovice, district Třebíč*

Near the parish church at Velký Újezd near Moravské Budějovice experimental geophysical measurement was carried out in 1980 for the Faculty of Arts, Masaryk University, Brno (further FA MU) whose purpose was to find the foundations of an assumed Romanesque rotunda in the area of the present cemetery. For solving this task, the SRP method was used with a double depth incidence (A1M1N1B, A2,5M1N2,5B - m) which, however, could be employed, with respect to the above situation, only at a limited space - two pairs of perpendicular profiles (Hašek - Měřínský 1991: 164).

From the data obtained it followed that in the studied area there exist relatively conspicuous changes in the sizes of specific resistances fluctuating in the interval of 90 to 250 ohmm. This is probably due to the above anthropogenic activity or by relics of stone masonry. The most marked zones of increased resistances that could locate the relics of buildings were suggested for archaeological verification. By probes carried out on the basis of geophysical measurements in 1980 the Art History Section of the Department of Art History, FA MU, Brno succeeded in one of three probes to discover remains of masonry, which, together with earlier written reports, contributed to the specification of the location of the sacral object and the knowledge of its shape. It was a rotunda with a cylindrical nave, evidently a horseshoe-shaped apse and at the elongated western side with a prism-shaped tower, maybe with a gallery. The whole formation can be dated only roughly to the end of the 12th to the 1st third of the 13th centuries (Kudělka et al. 1982-83: 86-87).

*Znojmo, district Znojmo*

The urban Gothic parish church of St. Nicholas at Znojmo is characterized by a three-nave hall with a long choir. The impetus for building it was probably the damage or destruction of the original, late Romanesque, probably one-nave church in the fire of the town in 1335.

The objective of the measurement by the DEMP method (h = 1.5 m, h = 3-5 m) and subsequently also with a georadar (Hašek-Kovárník 1996) was to find the course of the relics of the foundation masonry from the original building and the possible position of the crypt.

The results of geophysical works indicated that the foundation masonry of an older building (also in the negative imprint) is reflected by increased values of specific resistances with marked reflections of electromagnetic waves, particularly at the inner side of the circumferential masonry of the choir of the present object (Fig. 97). Its eastern limitation might be either in the shape of the apse or possibly rectangular. The object continues further into the places of the main nave, where it passes in parallel with a row of pillars. Its overall length could not, however, be established by the extent of the measurement (see Fig. 98).

The assumed position of the crypt below the church floor was not located unambiguously by the DEMP method. This can be due to the fact that it was filled with material of the same physical properties as has its surrounding environment. In the place of two shallow graves (Fig. 98) situated in front of the choir and covered with slabs a wide anomalous zone of increased resistances was measured. It is assumed that the graves were additionally cemented or tombstones set into the floor, or that there was only one southern tomb linked up with a possible filled crypt in the space of the choir of the original object. The GPR measurement in those places (Figs. 97 and 98) demonstrated the probable position of a half-filled cavity of dimensions of about 6 x 6 m.

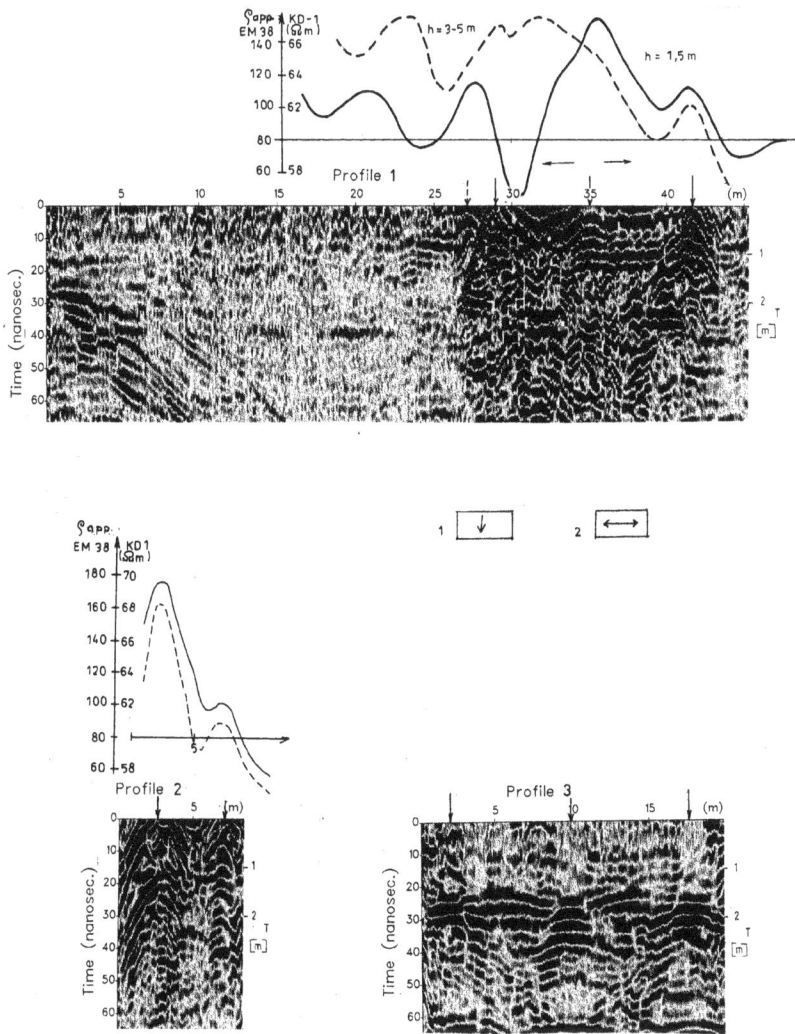

Fig. 97. 1- masonry relics, 2- areal inhomogenity

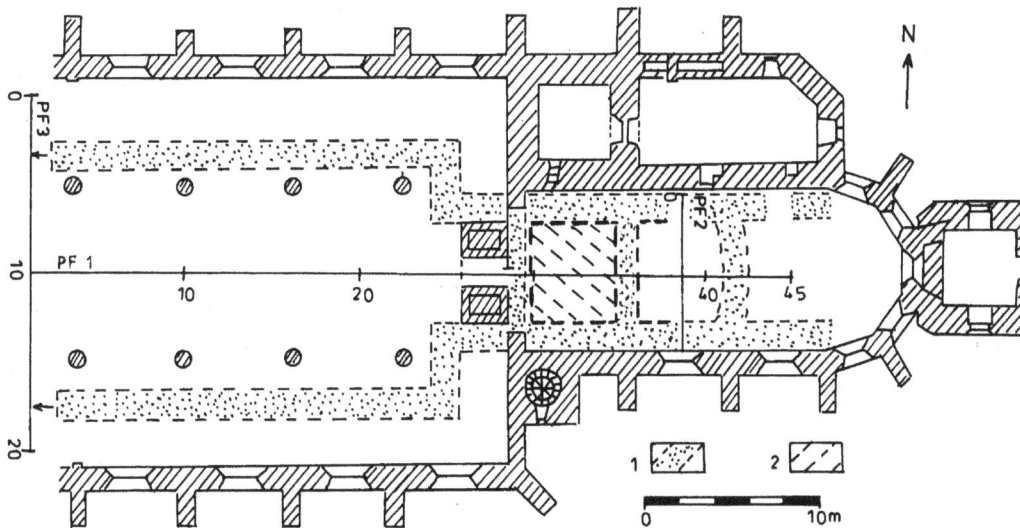

Fig. 98. 1- interpreted foundation masonry relics, 2- semiburied tomb

### 4.3.2.2. Seats of Holy Orders

*Doubravník, district Žd'ár nad Sázavou*

Geophysical works carried out in the space of the late Gothic church (1535-1557) in the community of Doubravník were to find the possible existence of the foundation masonry from the original parish church which had stood at the above place before 1208, and to which also a convent of Augustinian nuns was joined which was pulled down by the Hussites in 1423 (Oharek 1923). In the present church, besides the tombs of the Pernštejn and the Liechtenstein families also smaller 15th-16th century tombs are situated, and in 1867 the tomb of the Mitrovské family was built at it.

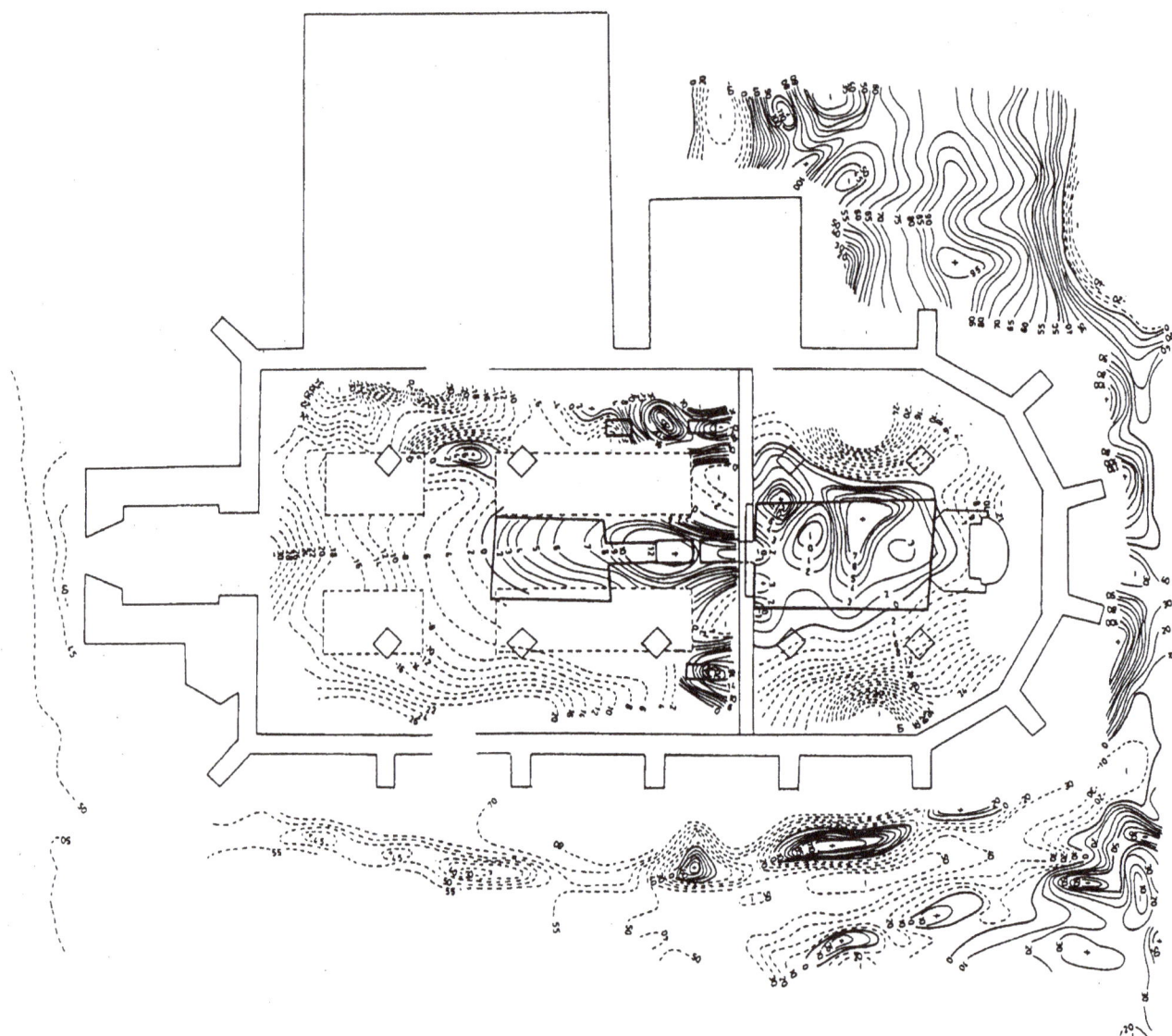

***Fig. 99.*** *Doubravník, district Žd'ár nad Sázavou. General $\rho_{DEMP}$ isolines (h= 3-5 m).*

A comprehensive processing of the DEMP measurements by apparatus with two depth ranges (h = 1.5 m and h = 3 to 5 m) indicated the course of the relics of foundation masonry of probably the original church, in the choir approximately lining the walls of the Pernštejn tomb and partly continuing even to the space of the northern and the southern side naves (Fig. 99). The western termination of this building, possible also with a spire (?) is assumed near the entrance to the Mitrovský tomb. The overall east-west length of the building can be up to 22 m, the width about 11 m (Hašek-Unger 1994: 30-43). The positions of the Pernštejn tombs below the choir and the Liechtenstein ones in the main nave appeared very distinct in the geophysical picture. Very interesting is a linearly oriented zone of increased resistances starting from the northeast corner of the Pernštejn tomb and continuing in the northern direction even outside the object of the church (see Fig. 99). Besides the building and geological affairs even the effect of a possible corridor (?) cannot be excluded. Some conclusions of the geophysical interpretation have only an informative character. It will be necessary to complete them by archaeological investigation.

*Jemnice, district Třebič*

The objective of geophysical works (Hašek 1993) carried out in places of the former Franciscan monastery was to find the possible existence of an older sacral building below the paving of the present Gothic church of St. Vitus, dating back to the 15th century, or even the position of the foundation masonry of the extinct monastery in its close neighbourhood.

From the results of DEMP measurements (h = 1.5 m, h = 3-5 m), in the space of the above building several areal zones were discovered as well as a narrower, linearly oriented zone of increased resistances that might locate the positions of later graves or tombs, the linear zone of higher resistances indicating, with respect to its direction, also the possible occurrence of masonry relics of an older stone building of circular ground plan (Fig. 100) (Hašek - Kovárník 1996: 65-88).

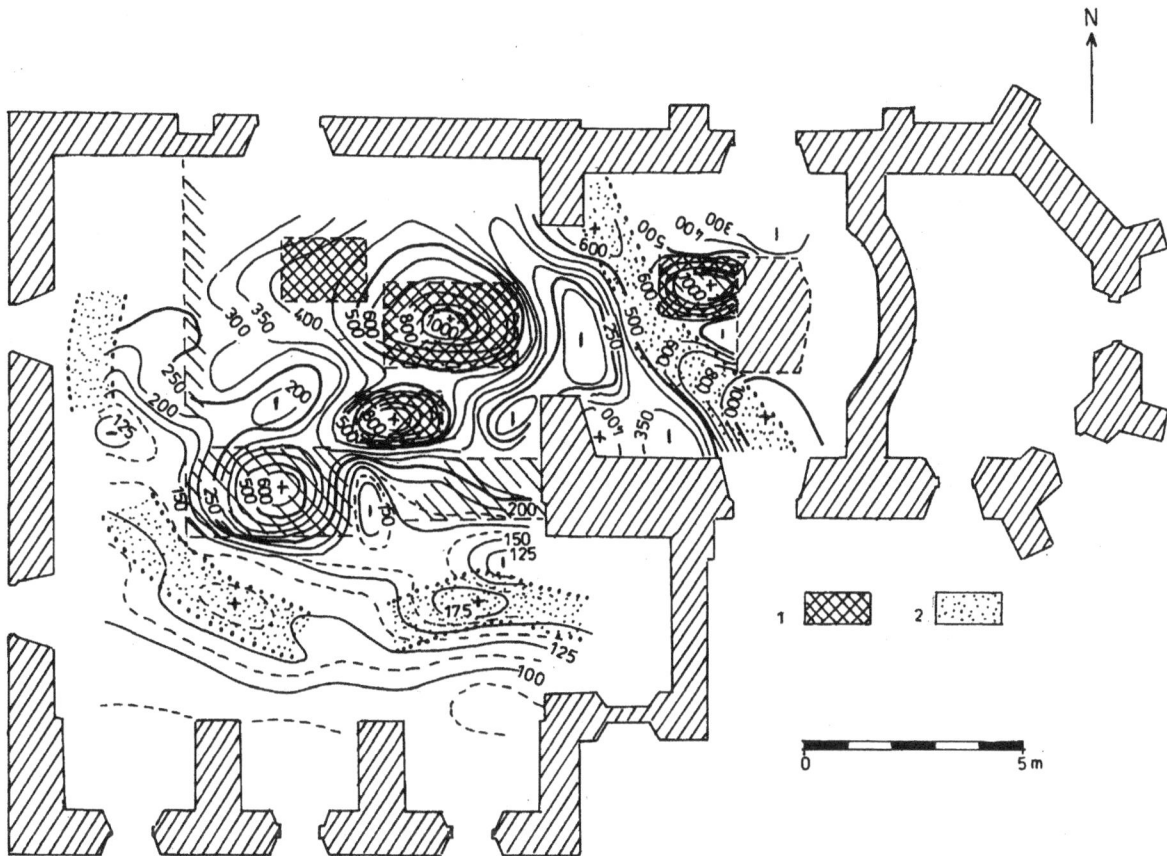

*Fig. 100.1- interpreted graves, 2- relics of foundation masonry*

According to the results of an experimental probe situated to the narrower resistance anomaly an object was found, recessed into the rock underlier, with remains of mortar affixes, which confirmed the geophysical interpretation assumption about the probable position of an older building in places of the Gothic church of St. Vitus.

The results of the preliminary geophysical measurement carried out outside the above building indicated the course of the relics of the masonry of the extinct monastery.

*Předklášteří near Tišnov, district Brno-country*

Geophysical works carried out in the area of the Cistercian convent Porta Coeli were to delimit the position and the ground plan of the chapel of St. Catherine dating back to the 13th century, assumed to be in the NW part of the court (Belcredi 1993: 323-340).

Data from the SOP and DEMP methods (h = 3-5 m), processed into a map of isolines $\rho_{DEMP}$ (Fig. 101) indicated approximately two narrow, linearly oriented regions of increased resistances with the axis in the NE - SW direction, the northern one being impaired by another zone of increased resistances, oriented in the direction NNE - SSW (Hašek-Kovárník 1996: 65-88). Towards the northeast the above two main anomalous belts become narrower and their gradual joining (closing of the choir) can be assumed.

Into anomalous places found by measurement an area excavation was situated which subsequently found relics of combined and brick foundation masonry of a building of overall dimensions of 16 x 9 m, and the trace of engineering networks. The exposed graves inside the building (excepting the mensa) could not be located by measurement due to their size, depth, and, last but not least, their physical markedness.

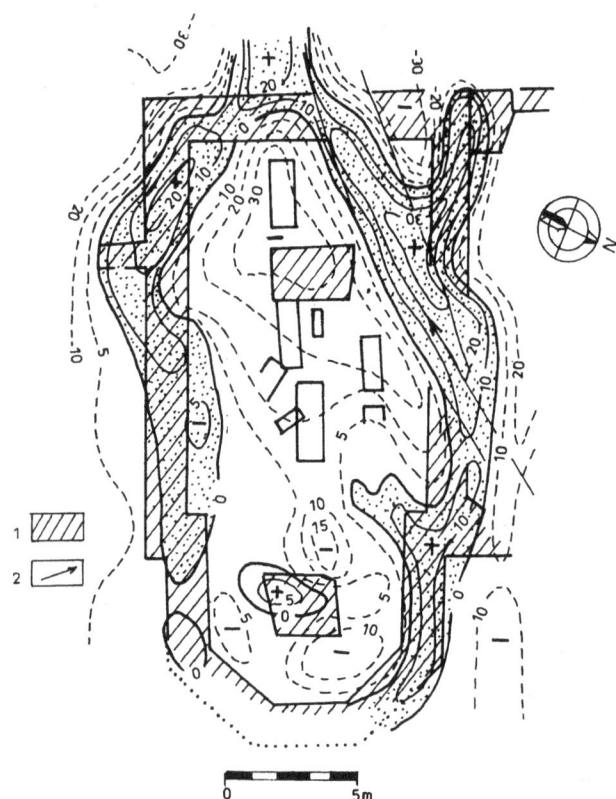

*Fig. 101.* 1- exposed object, 2- engineering networks

## 4.4. Investigation of Towns

A marked phenomenon of the comprehensive medievalistic study are medieval towns arising in the territory of Moravia either as agglomerations at main Moravian castles of the early Middle Ages and the seats of appanage princes (Brno, Olomouc, Znojmo) or founded from the 13th century by Premyslide rulers. An important element for the importance and position in both royal and liege towns was the existence of municipal walls. Royal towns had been mostly fortified before, whereas liege towns, due to the overall bonds to the owner, were fortified rather sporadically and later (Razím 1995: 9-22). Besides the fortification, also the study of the town house and further objects (such as sumps) have a great illustrating value. Archaeogeophysical prospection performed was oriented on the above moments.

### 4.4.1. Urban Historical Cores

#### 4.4.1.1. Fortification

*Jihlava, district Jihlava*

The objective of geophysical works, carried out for purposes of the rescue geophysical investigation in the historical core of Jihlava (Hašek-Mitrenga 1992) was to follow up the course of relics of foundation masonry of medieval fortification dating back to the 13th to 15th centuries, or of further possible near-surface inhomogeneities in its surroundings.

The results of geophysical measurements reflected in maps of isolines $\rho_{DEMP}$ and grad. $T_z$ (Fig. 102) indicated the course of two fortification walls of different widths turning from the NNW - SSE to the WNW - SES direction and of a moat.

Relics of masonry can be considerably impaired in places or completely destroyed. From the map of grad $T_z$ (see Fig. 102) both considerably inhomogeneous material is assumed (stone, brick) and a thicker layer of the destruction debris in the vicinity of the moat wall. At the eastern margin of the area processed it is possible, according to the results of geophysics, to separate the trace of a sewer approximately in the NNE - SSW direction (Šedo-Zatloukal 1993). The data of the geophysical processing are in accord with the investigation.

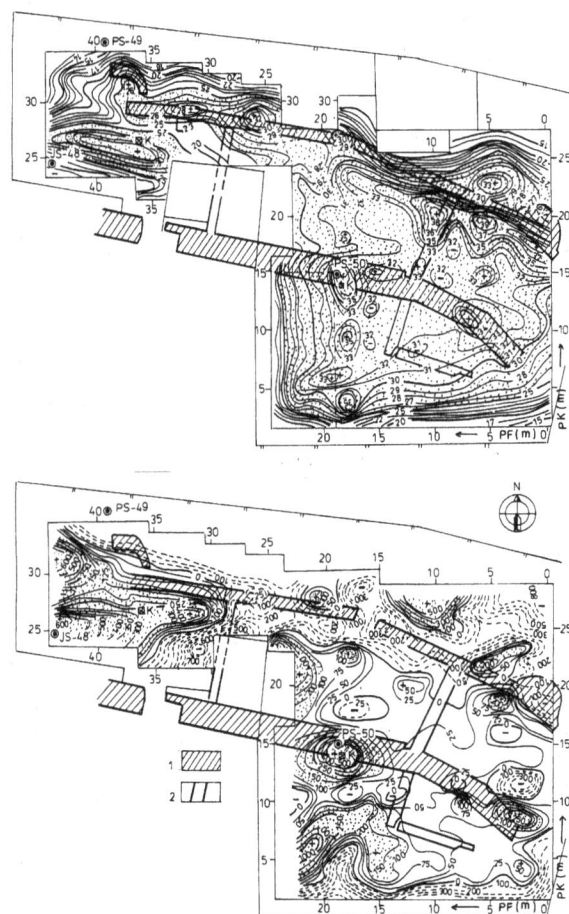

*Fig. 102.* 1- exposed fortification, wall, 2- drainage

*Nový Jičín, district Nový Jičín*

The objective of geophysical and subsequent verifying boring works carried out by magnetometry and the DEMP method in front of the main fortification wall at Nový Jičín (Hašek et al. 1995) was to yield detailed information about the position and the course of the town fortification dating back to the 14th and the 15th century and even the later building up of the studied area of the size of about 2400 m² for the purposeful situation of the subsequent rescue archaeological investigation of the District Museum of Local History at Nový Jičín.

From the results of the processing of magnetometric measurement in the area studied a number of local isometric or linearly oriented positive anomalies $T_z$ was found whose sources will probably be objects of modern building up activity of dimensions of about 2 x 2 m to 3 x 8 m located near the surface of the field. In comparison with the results or boring works they are above all relics of brick masonry (æ =

0.994 to 8.693 x $10^{-3}$ u.SI), partly collapsed cellars (æ of the filling = 39.08 x $10^{-3}$ u.SI), destruction layers situations of prefabricated parts, traces of engineering networks, etc.

Similar properties are exhibited - in correlation with magnetometry - also by data from the DEMP method. Increased values of specific resistances (h = 1.5 - 2 m) locating the positions of stone (sandstone) and brick masonry or large concentrations of this building material, cellars, extent of gravel-sand near the surface as against the reduced

values (h = 3 - 5 m) which correspond with the distribution of clay-loamy soils.

The results of the employed methods further indicated the course of a parallel stone sewer, width about 1.5 m and depth 2 m, at the distance of about 6 - 8 m from the main wall, the position of a moat, 10 - 15 m wide and more than 6 m deep, filled with dark mud or clay.

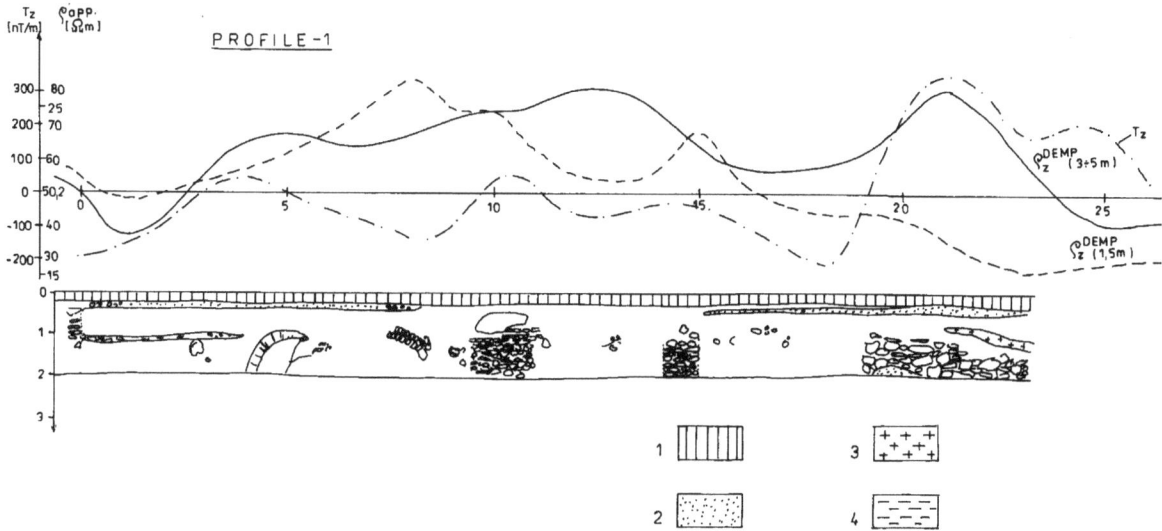

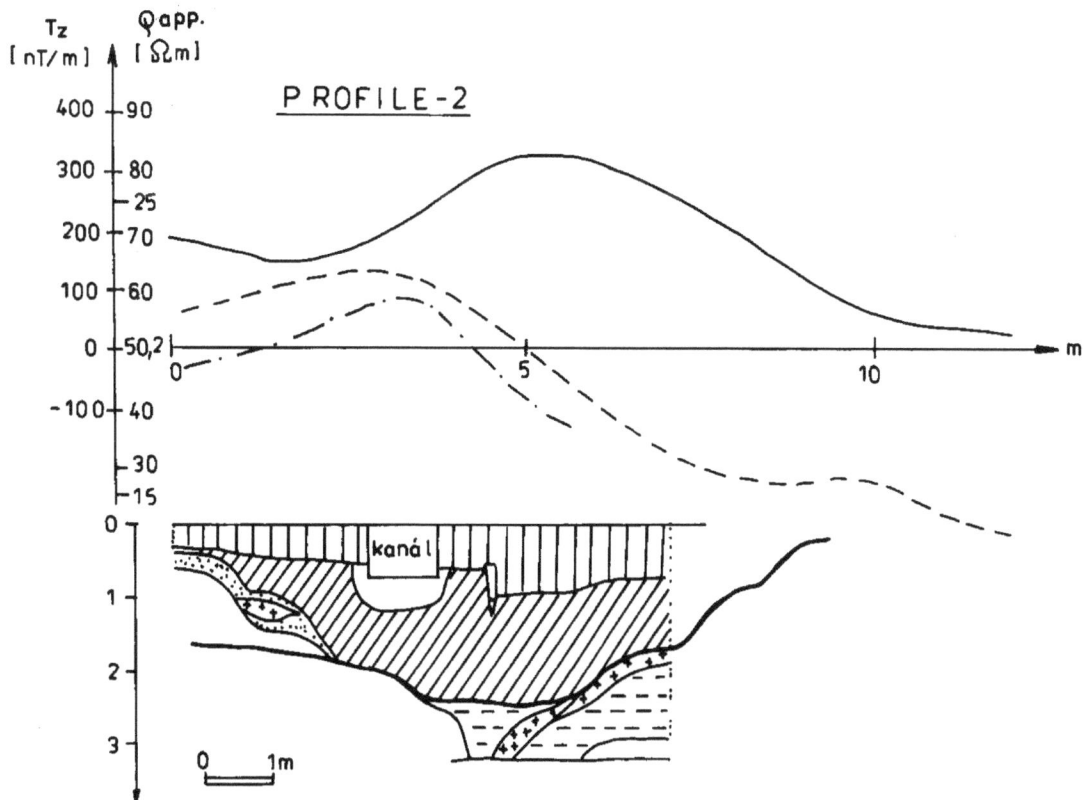

*Fig. 103. 1- made-up ground, 2- gray green clay (dark shade), 3- plaster, 4- schist*

The performed archaeological probe confirmed fully the conclusions of the geophysical interpretation (Fig. 103). From the methodological point of view, the combination of geophysical and boring works in this type of tasks proved very useful. The determination of the overall archaeological and geological characteristic of the studied area helps considerably to site excavation works.

### 4.4.1.2. Houses and Farming Objects

*Olomouc, district Olomouc*

In connection with the reconstruction of the basement spaces of the building of the Local History Museum at Denisova

Street No. 30, situated in the historical part of Olomouc, experimental geophysical measurement by the DEMP method was carried out in the above places with apparatus of different depth ranges, whose task it was to find the assumed relics of older building activity, preceding the building of the present object (Hašek-Bachratý-Tomešek 1993). The above building is situated at the very frontier of two urban units, the western part called Předhradí (suburbium) and the proper medieval town which were separated by the municipal wall running directly under the complex of originally Jesuit buildings built gradually from the latter half of the 17th to the early 18th centuries (Tymonová 1993).

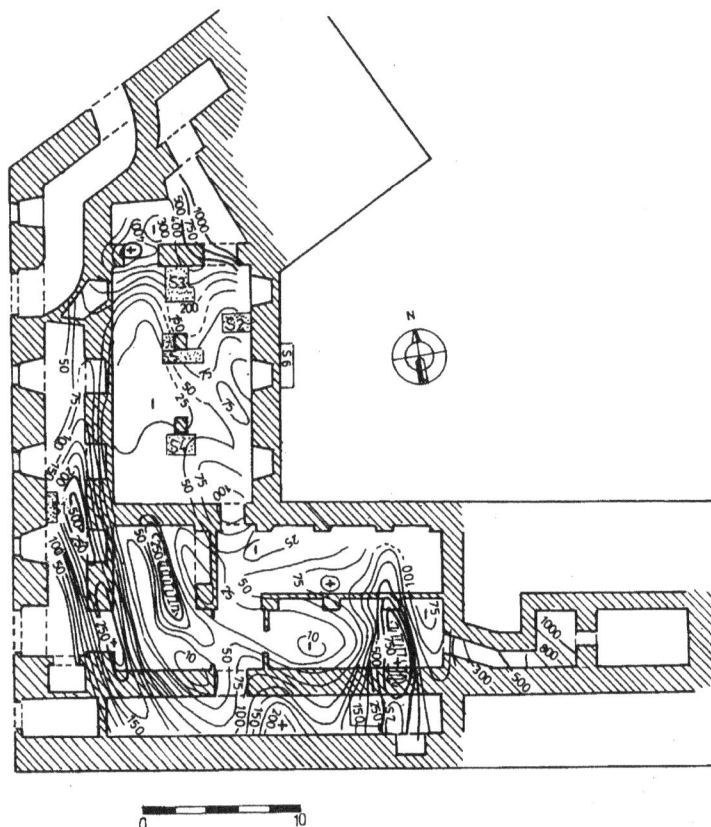

*Fig. 104. Olomouc, Denis Street. $\rho_{DEMP}$ isolines and situation of archaeological research.*

From the map of isolines $\rho_{DEMP}$ (Fig. 104) a linearly oriented zone of increased resistances was found along the present western circumferential wall, in the NNW - SSE direction which, as verified by probe S-1, is constituted by a compact dark red layer of clay-sandy loam with stones, probably filling some moat-like formation. The find material can roughly be dated to the 13th century, the dark grey to black filling is connected with the young and late ringwall cultural layer trenching there from Republiky Square there (Tymonová 1993).

Another linear zone of increased specific resistances was come across approximately in the central part of the area studied. Its direction orientation is NW - SE (see Fig. 104). By archaeological verification (probe S-5) a block of masonry was located there passing directly in a slanting way through the probe, approximately in the N - S direction and

rising to 40 - 45 cm from the present surface. The resulting foundation masonry consisted of irregular stones connected with yellowbrown calcareous mortar which, at the depth of 60 - 70 cm bore traces of burning. The same also concerned the continuing block in the place of NW widening (Tymonová 1993).

In the southern habitation rectangular room on the right side of the entrance in the direction of Universitní Street an approximately isometric region of increased resistances was discovered, accompanied by a slight depression in the floor of concrete tiles. Archaeological excavation (probe S-7) demonstrated a lined sump to the depth of 1.90 m under the relics of rotten and partly burned planks. Its inside was filled to the depth of 60 - 70 cm with black compact and towards the base also muddy loam with a considerable amount of pottery material found first at the depth of 50 - 60 cm from the surface. The material obtained from the content of the sump can be dated to the end of the 15th to the 17th centuries, which documents its time of duration. The position of the sump at the western side of the assumed town wall indicates the fact that earlier patrician houses may have been there (Tymonová 1993).

*Tábor, district Tábor*

In connection with the reconstruction of a house dating back to the latter half of the 16th century in the historical part of Tábor, the task of measurement by the radiolocating method (Hašek - Krajíc - Tomešek 1996) concentrated on the verification of the course of the cellar under the present communication and further on the existence of a possible entrance into the above object from a room of building No. 54 in the first basement.

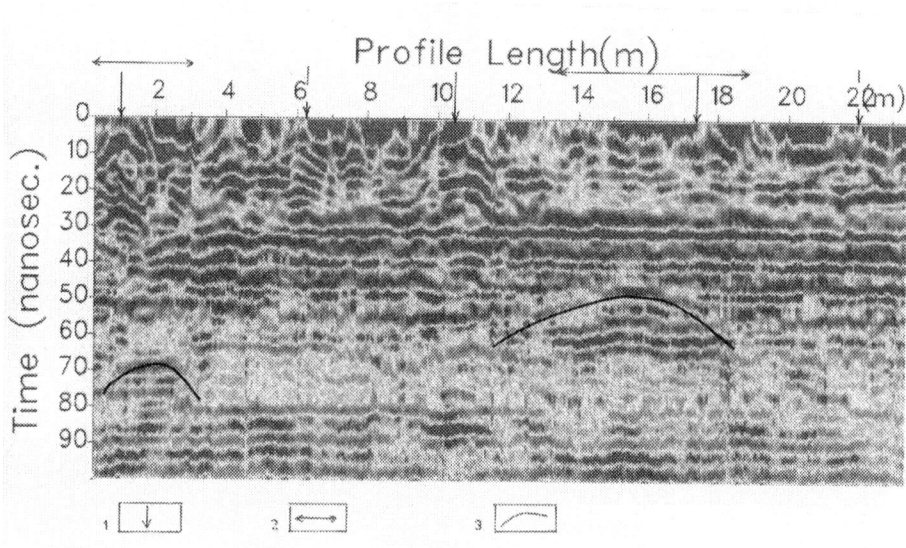

*Fig. 105. 1- local inhomogenity, 2- areal inhomogenity, 3- interface: cover formation-rock skelet*

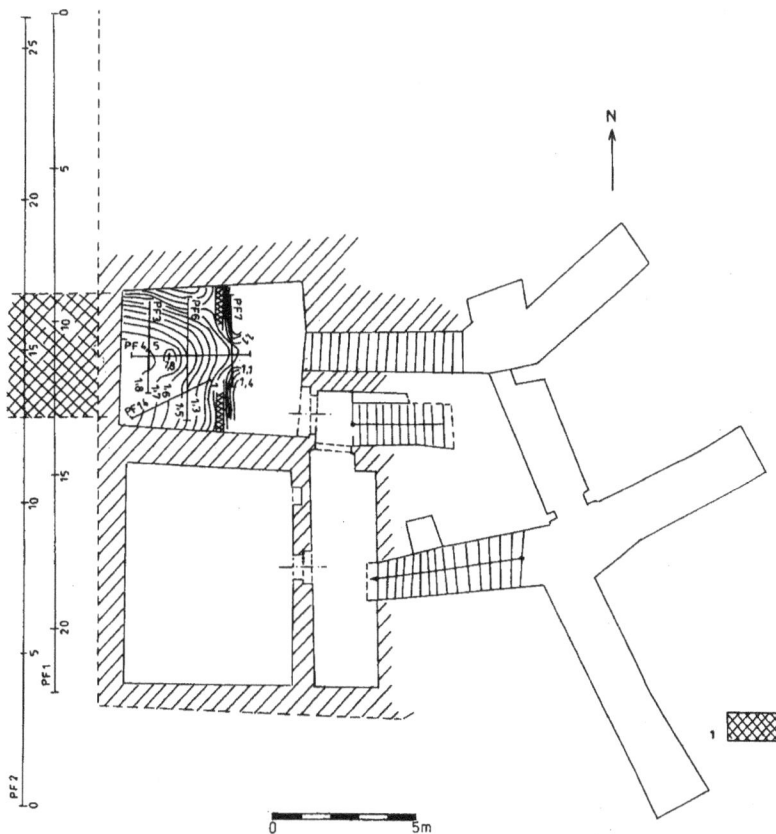

*Fig. 106. 1- interpreted cellar*

From the results of GPR, in the area of Arbeiterova ulice Street, using an areal of 50 and 200 MHz, a number of anomalous elements was found (Fig. 105) that might be assigned both to the indication of cavities, and traces of engineering networks. Relatively marked amplitudes of the reflected electromagnetic waves are at the time of 30 ns, i.e. approximately in the depth of 2.0 m. The relief of the Tábor syenite is interpreted there. At the picket of 13 to 17 m (see Fig. 105) it is possible to locate a half-filled cellar extending to the above places probably from the space of the first basement of house No. 54. The approximate depth of the ceiling of the inhomogeneity is about 3.5 m, the width 4 - 4.5 m. Another small anomaly (cavity ?) can be separated in the surroundings of picket of 2 m (see Fig. 105). According to the indication in time in the vicinity of 70 ns its depth is assumed to be about 4.5 m.

In the first basement of object No. 54 a relatively large thickness of the made-up ground was located by the measurement in the central part of the room, varying from 1.0 to 1.8 m in the direction of W - E (Fig. 106). The relatively largest increase can be observed in the surroundings of PF 5.5, approximately in places of the assumed entrance to the cellar discovered by GPR in the space of the communication. At the northern and eastern margins of the room there is an evident effect of both rocks and the interference of waving from the existing masonry. The results of interpretation are fully in agreement with archaeological investigation.

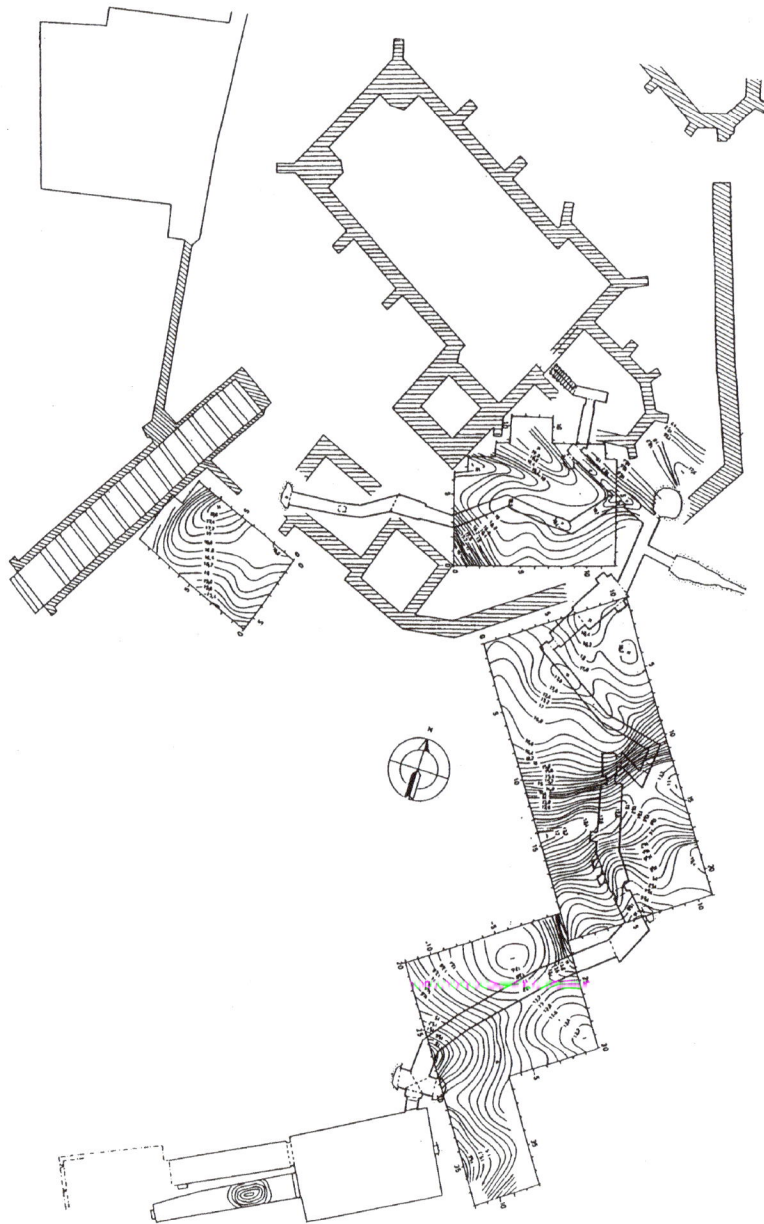

**Fig. 107** *Kurdìjov, district Bøeclav. $\rho_{DEMP}$ isolines and map of underground corridors.*

### 4.4.1.3. Cellars, Corridors, Unvaulted Cellars

*Kurdějov, district Břeclav*

Geophysical works in the wide space of the fortified church at Kurdějov were to verify the possibilities of the DEMP method for the location of escape corridors dating to the turn of the 16th and the 17th centuries (Unger 1987: 5-19) of the height of 1.5 - 1.95 m and the width of 0.7 - 1.6 m dug in loess and lined with brick, stretching at the depth of h = 2 to 5 m between the cellar of the present pub and the above church building (Hašek - Unger 1994: 30-43).

The results of geophysical measurements agree very well with data of archaeological investigation (Fig. 107). The traces of underground corridors appear clearly from geophysical data as narrow zones of increased specific resistances, particularly in the area of the church, in the SW sector of the area of interest they indicated among others also the possible continuation of the corridor in the slope behind the collapsed part with its deflection from its original direction (E - W) to the NE - SW orientation.

Outside the object fortification the location of the corridor is also very conspicuous, even though in the middle part there is a certain distortion of its course, due to probably the position of the aqueous extenuated zone (direction NE - SW) and/or by the lithological change in the loesses. In the surroundings of PF 10 PK 20 and 35 m the results of the measurement are affected by the effect of a cellar into which the followed corridor opens.

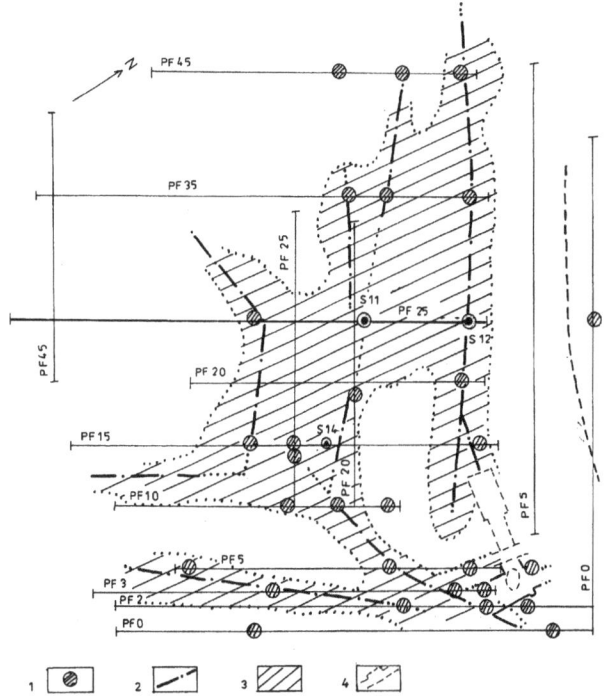

*Fig. 108 1- inhomogenity indicated by GPR, 2- axis of non-conductive zones, 3- area of lower conductivity, 4- cellar*

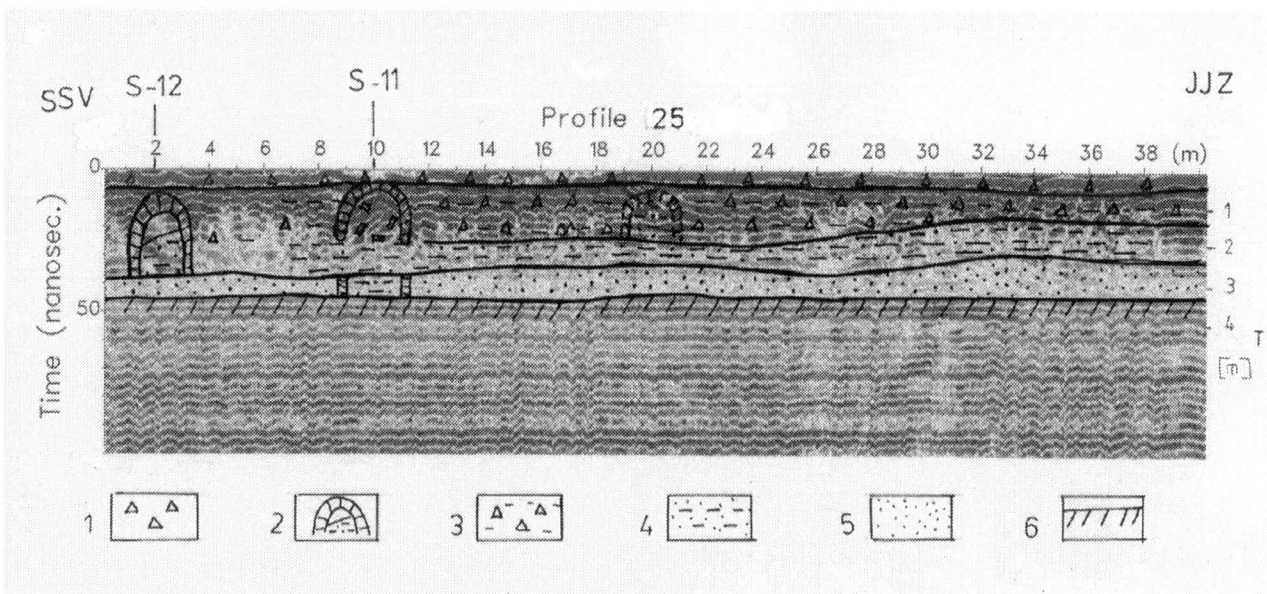

*Fig. 109 1- made-up ground (limestone, bricks), 2- cavities with brick filling and/or with clay and sand re, 3- soil with limestone fragments, 4- sandy soil, 5- medium-grained sand, 6- argillaceous sandy clay*

*Mikulov in Moravia, district Břeclav*

In the space of the planned parking lot in the area near A. Muchy Street, passing through places of the former historical built-up area of Mikulov in its northeastern sector the paving broke through into underground spaces. In the caving the found roof of the cavity was in the depth of 0.6 m. The continuation was found by investigation approximately in the direction of A. Muchy Street with several crossings, mostly partly collapsed or walled-in. The vaults are made of brick, the age of the found cellars was estimated to the 19th and the beginning of the 20th centuries (Unger 1996).

The task of the measurement by georadar and DEMP (Dostál - Hašek - Tomešek 1996) was to follow the position and further course of cellar spaces and verify their possible existence by boring works.

The results of the prospection on the studied area of about 55 x 45 m proved the position of linearly oriented zones of reduced conductivities accompanied by multiple reflections of electromagnetic waves from GPR (Fig. 108) particularly in the southern and the eastern sectors. The boring investigation indicated that those zones were indications of partly collapsed cellars (S-11, S-12 see Fig. 109) and in the surroundings of S-14 also relics of masonry from older buildings. The continuation of underground spaces behind the caving-in of the paving is assumed in the NW - SE direction. But two parallel complicated systems in the vicinity of probes S-11 and S-12 cannot be excluded. The interpreted depths of vaults are about 0.7 to 1.1 m. The probable effect of the relics of foundation masonry is expected at the southeastern margin of the area of interest, which is also in accord with the data of historical topographic maps.

*Rabmühle near Roding, Germany*

In places of the well-known underground spaces - an unvaulted cellar near Rabmühle geophysical measurement by the DEMP method was carried out with the objective of mapping the course of the above inhomogeneities consisting, with the exception of the entrance part, by prevailingly unlined corridors of different heights and widths into the depths of 5 to 6 m (Hašek-Unger 1994a: 27-29).

From the drawn map of isolines ρDEMP (Fig. 110) it follows that the mostly narrow zones of increased, in places also reduced (the surroundings of the well) magnitudes of specific resistances locate relatively reliably the position and course of known (collapsed as well as uncollapsed) underground corridors. The more extensive regions of increased resistances indicate, according to their character, the possibility of existence either of further continuation of hitherto unknown spaces in the NE - SW direction and/or also another entrance into this system (PF 0 PK 15 m). No further anomalous elements that could be allocated to the possible course of hitherto uninvestigated corridors have been demonstrated by measurement.

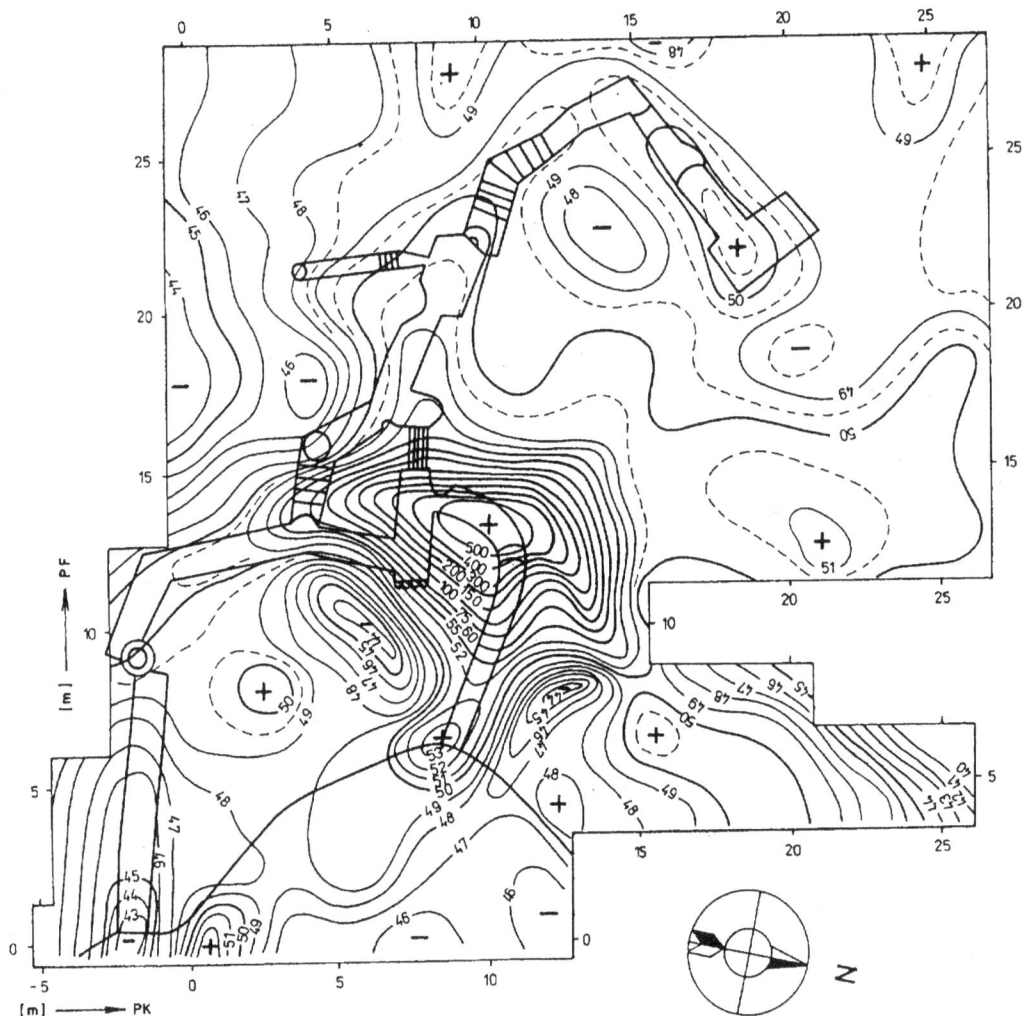

*Fig. 110. Rabmühle near Roding, Germany. ρDEMP isolines and their comparison to the course of underground corridor - loch.*

## 4.5.Investigation of Burial Grounds

In looking for sepulchral memorials, field archaeological practice in the CR has been dependent above all on accidental finds of graves at different interventions into the field, and that is why the location of burial grounds, barrow fields and the determination of the extent or the character of the necropolis is one of the first-rate tasks of geophysical prospection. The starting area for obtaining experience are above all burial grounds with rich iron inventory, mainly from the La Téne period and particularly from the medium and late Ringwall Period (Hašek - Měřínský 1991: 135-140).

A special position in solving those issues is that of tombs. In this country it is above all the location of crypts as part of the medieval sacral architecture, unlike the desert territory near Abussir (Egypt) where, besides royal tombs - pyramids (structures built of limestone blocks with a granite coat) also non-royal tombs were built, called mastabae. They consist of an overground part (enclosing wall, chapel, etc.) and an underground one (shafts, sarcophagus chambers) of limestone, dried brick, etc. The two kinds of structures can be included into the period of the 5th dynasty of the Old Empire (Hašek - Obr - Verner 1988: 5-47). In the late period also further structures of the burial character were built, quite different from the above ones. They are shaft graves consisting of a system of interconnected deep shafts which, after the deposition of the sarcophagus, were covered in with

sand. The overground part of the object, such as the enclosure, the lining of the shafts, etc. were prevailingly built of limestone blocks, often combined with dried brick.

In the archaeogeophysical prospection of sepulchral memorials it is possible to employ both magnetometry and geoelectric methods (RP, DEMP), or their combination. In locating crypts, besides georadar also microgravimetry is effective (Hašek - Měřínský 1991: 141-144).

### 4.5.1. Flat Inhumation Burial Grounds

*Velešovice, district Vyškov*

In the area of motorway construction near the community of Velešovice a burial ground of the Corded Ware Culture (2300 - 2000 B.C.) was discovered by the earth works and the subsequent rescue archaeological investigation of the AI CSAS in Brno as well as settlement objects of the Únětice Culture. Magnetometric measurements carried out in 1988 on the area of 100 x 50 m were to find out the positions of the individual graves and the extent of the settlement in the southern direction from the hitherto investigated area (Hašek-Měřínský 1989: 103-151). From the map of isanomales ΔT (Fig. 111) it was possible to separate, besides a conspicuous linear anomaly due to the gas pipeline, a number of approximately isometric anomalies ΔT which were recommended to archaeological verification.

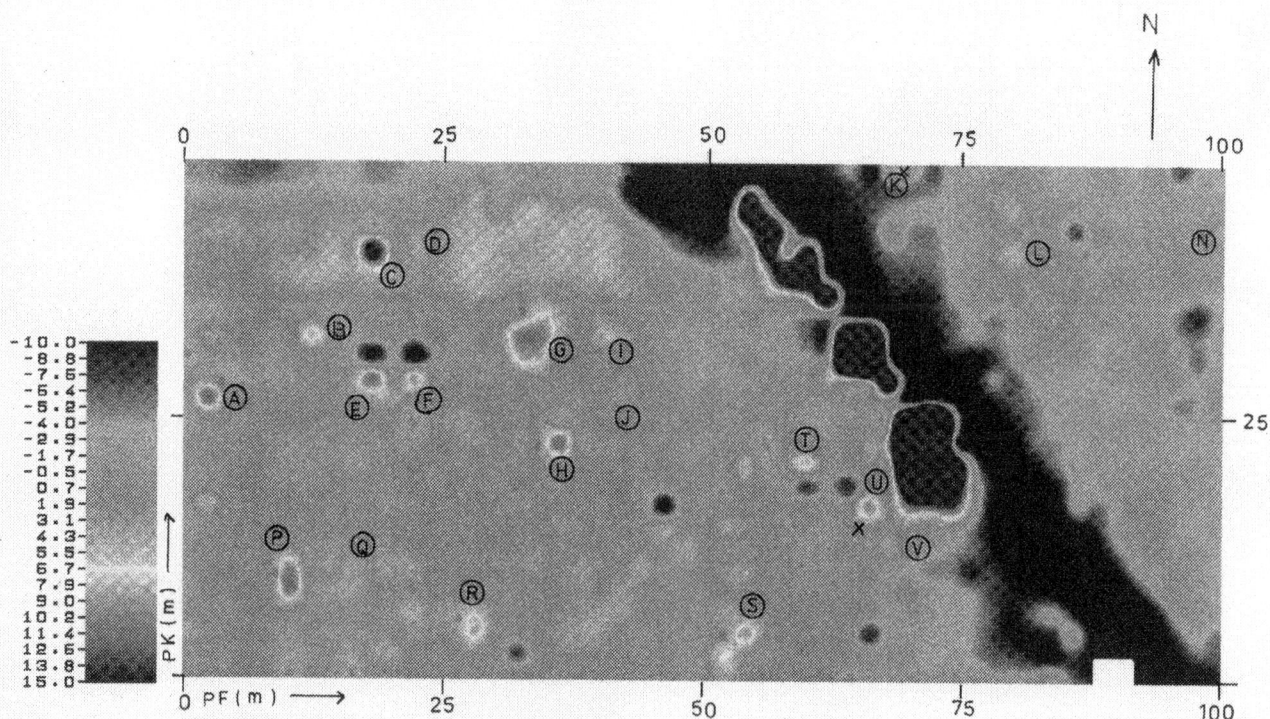

*Fig. 111. Velešovice, district Vyškov. Map of ΔT isanomales with marked archaeological objects. 1- graves, 2- settlement objects*

The excavation works carried out and the pedological probing demonstrated the fact that their sources were settlement objects of the Únětice Culture (anomalies T denoted as A, B, G, H, P, R, S), the filling of the Corded Ware Culture graves of the dimension of roughly 200 x 220 cm and the depth of 80 cm (K, V) as well as the cultural layer unrecessed into the substrate (L, M). From the data of

geophysical measurement and the find situation it is assumed that the burial ground of the Corded Ware Culture penetrates into the studied area only by its southern margin. Its possible ending there is expected, contrary to the relatively intense settlement of the whole area.

*Velké Bílovice, district Břeclav*

A burial ground with 71 + 2 graves dating back to the middle Ringwall Period discovered during earth works in 1975 and investigated by the AI CSAS in Brno in 1976 was situated on an elevated place of an elongated ridge at the foot of a slope with southeastern exposition. The slope had a dissected microrelief in those places which, in the youngest geological period, was due to the accumulation of loess and rainwash sediments. Originally the area of the necropolis was on a sort of peninsula extending into the inundation, and during the investigation the surface humus horizon of the thickness of 40 to 90 cm was removed. The substrate in which grave fillings were evident by the different colouration, consisted of Neogene sands, in places mixed with loam (Měřínský 1984a: 41-44). In the area of the burial ground large empty spaces were found in which it was impossible to exclude the fact that there might have been graves there with fillings undistinguishable from the substrate. Along the northwestern to western and the southwestern to southern margin of the necropolis, in a depression under deposits of as much as 2 m thick of the surface soil, a belt of the fluvial plain gley soil was found (Měřínský 1984a: 44, Fig. 4 on p. 43; Ludikovský - Hašek 1978: 137).

Archaeophysical prospection carried out in 1986 was to verify the possibility of finding objects of the grave goods before opening the grave pit and to determine whether in the case of the above belt of the fluvial plain gley soil of the width of 3 - 5.5 m it is not a moat (Hašek - Ludikovský 1977a: 112-113).

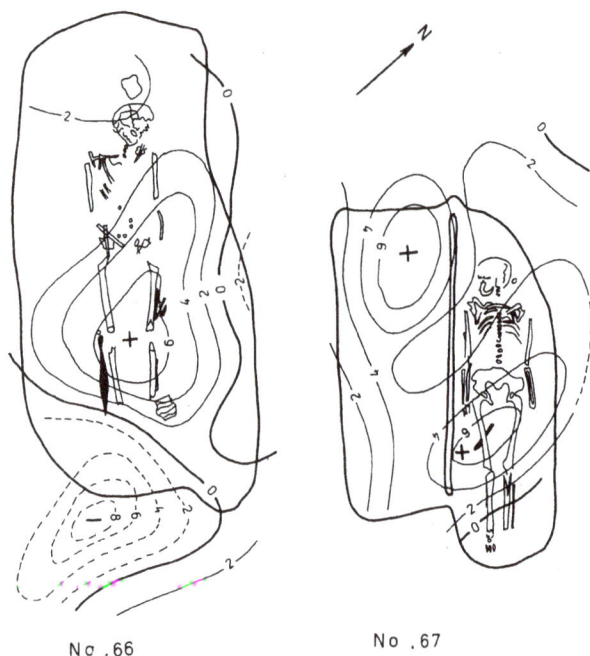

No.66    No.67

**Fig. 112.** *Velké Bílovice, district Břeclav. Map of ΔT isanomales and siruation of the archaeological research.*

By a detailed magnetometric measurement carried out at the areas of interest in the network of 0.5 x 0.5 m three outlined grave pits Nos. 66, 67 and object No. 5 (grave No. 71) were studied, whose filling differed in colour from the substrate. The outline of the rectangular grave No. 66 (Fig. 112) yielded by the characteristic of isanomales ΔT information about the overall shape of the grave pit. The increased value

of the local anomaly of the magnetic field (+ 8 nT) locates the position of Fe-objects - a spear, a knife and pottery. The irregular filling of grave No. 67 was interpreted as a double grave on the basis of the distribution of isanomales ΔT. The anomaly ΔT (+ 6 nT) indicated the presence of an Fe-object in the NE part. After removing the filling it was found that it was a grave in a niche containing a skeleton with a knife under the right femur and a bronze ear ring. The niche was separated from the grave shaft proper by a wooden partition (Měřínský 1985: 120-124, Figs. 41-42 on pp. 123-124).

In the third case irregular rectangular filling in the N corner of the burial ground was investigated, where the anomaly ΔT was + 4 nT. In the investigation it was denoted as object No. 5 and it was a well-like oval shaft 2.98 m deep, in which, at the depth of 245 to 275ţcm a human skeleton had been thrown in (grave No. 71) dated by pottery also into the Great Moravian period (Měřínský 1984a: 57, Fig. 7-9 on pp. 45-47; 1985: 120-125, Figs. 41-44 on pp. 123-126).

A shallow trough, as shown by two probes only 0.5-0.6 m deep at the margin of the burial ground, could be followed magnetometrically clearly along its whole length. It was a natural, originally at least seasonally watered depression through which a small water tributary to the brook was flowing (Měřínský 1984a: 44; cf. Hašek-Měřínský 1991: 137-138).

### 4.5.2. Barrow Fields

*Bohuslavice near Kyjov, district Hodonín*

Magnetometric measurement carried out in the space of the Slavonic barrow field dating back to the 9th century on areas of 95 x 50 m and 40 x 35 m in a network of 1 x 1 m was to map its extent and to find the possible positions of further objects. From the performed modelling of the effect of the barrow (Fig. 113) by means of values of magnetic susceptibility found on the walls of the exposed object (defence line - vallum - æ = 0.53 .$10^{-3}$ u.SI, the surrounding environment æ = 0.2 .$10^{-3}$ u.SI) it was found that the source of anomalies ΔT was the filling of the barrow, the grave pit only emphasizing its maximum.

In the map of anomalies ΔT (Fig. 114) the individual barrows are expressed by the positive isometric anomaly of the magnetic field of a really different size, either with one or with two partial maxima. In the case of barrows in which a secondary intervention has been made the positive anomaly ΔT is impaired in the peak part by a local isometric negative anomaly (see Fig. 114).

The linear negative anomaly ΔT at the southern side of the studied area is due to a shallow moat limiting the extent of the barrow field in this sector.

Geophysical measurement both demonstrated the existence of all barrows from the surface mapping and yielded some complementary data about further objects which are no longer evident in the field relief. This concerns above all the wider surrounding of barrow No. 15 (Fig. 114), where a linear anomaly ΔT was separated bordering the above object which might even correspond to the course of a shallow moat, further a number of extensive isometric anomalies ΔT probably linked up with further object of the funeral rite.

The cause of anomalies with two maxima (e.g. Nos. 30, 15, 19, 22, 11, 1, 2, 4) can be the positions of two grave pits in one object, inhomogeneities in the defence line - vallum and/or two small barrows in a close vicinity.

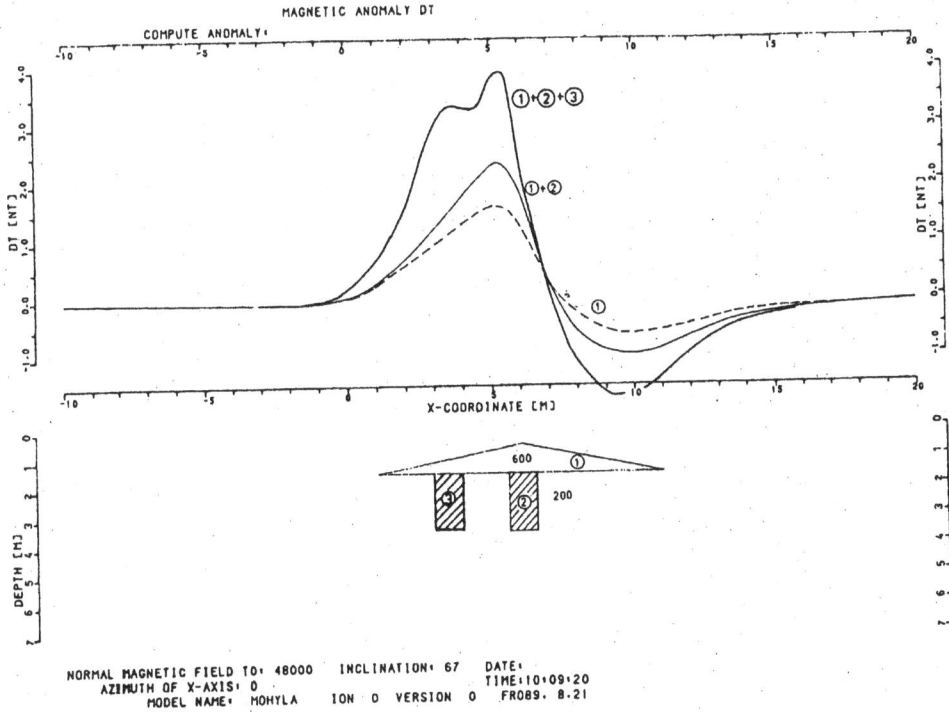

*Fig. 113. Bohuslavice near Kyjov, district Hodonín. Model of barrow effects.*

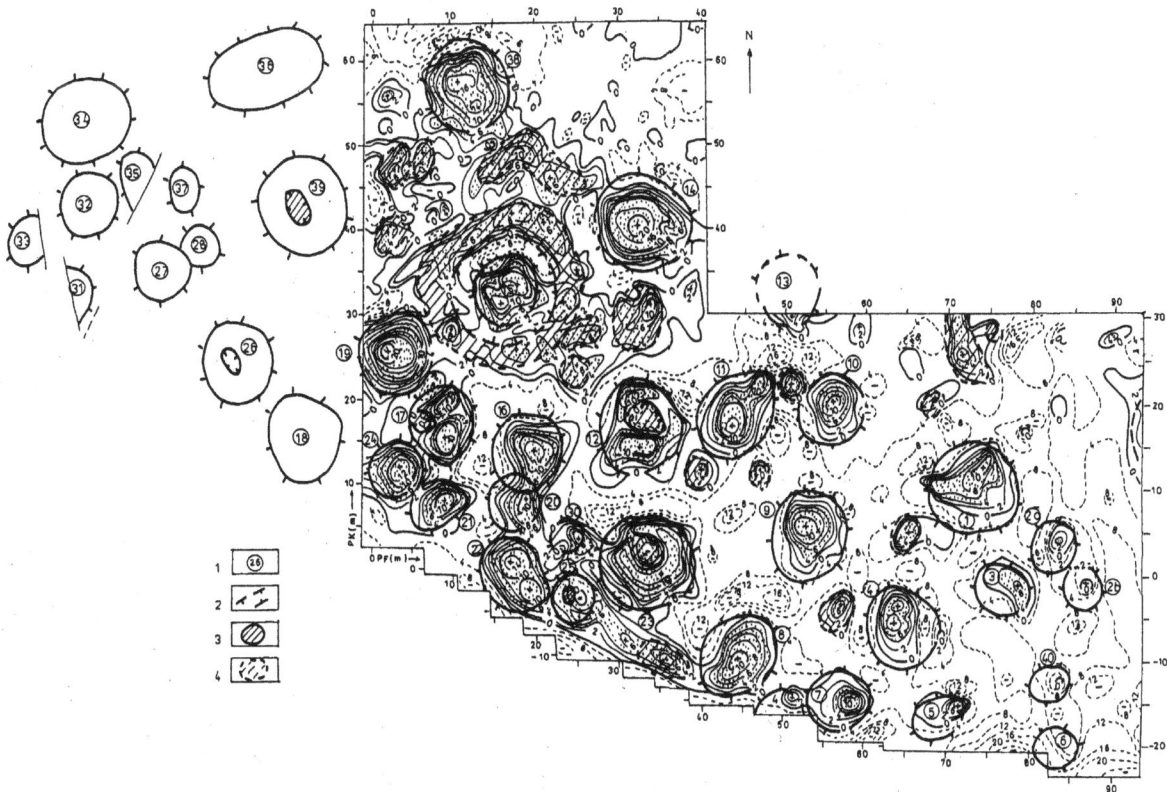

*Fig. 114. 1- object marking, 2- tumulus outline, 3- secondary intervention, 4- ditch*

## 4.5.3. Cremation Burial Grounds

### Podolí, district Brno

From 1974, the AI CSAS in Brno carried out a revision archaeological investigation at the cremation burial ground of the eponymous station of the Podolí phase of the Middle Danubian Urn Fields (Říhovský 1977).

The purpose of geophysical works in 1977 was to find the extent of the burial ground and/or the positions of the individual cremation graves. A detailed magnetometric measurement took place on an area of 15 x 50 m with the step of 1 x 1 m (Hašek et al. 1977: 39-42). The course of the magnetic field (see Fig. 115) is considerably affected by the effect of recent Fe-objects in the western and southern parts of the studied area (wire netting fence, etc.), only the central part can be utilized for determining the probable position of the individual studied objects. The results of the measurement permitted us to separate a number of isometric anomalies ΔT which can locate the positions of the individual cremation graves. Even the assumption cannot be excluded that some of the measured anomalies of the magnetic field are due to filled holes after fruit trees. For verifying probes the space denoted as A - A (Fig. 115) was suggested in which three anomalies ΔT appeared (+ 20 nT, + 25 nT, + 25 nT).

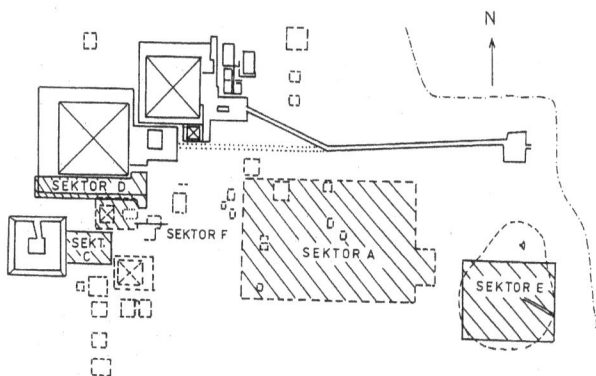

*Fig. 115. Podolí, district Brno. Map of ΔT isanomales and position of archaeological sound.*

The studied area A - A was exposed by probing (8 x 2 m) and from the three anomalies ΔT found two proved to be positive. Each contained simple cremation graves (large urns, small amphorae, a pot-like vessel, etc.) at the depth of about 100 cm below the present surface level (Říhovský 1980), the third, judging from a great amount of roots, probably corresponded to the filling of a hole after removing a tree (cf. Hašek - Měřínský 1991: 138-140).

## 4.5.4. Tombs

### 4.5.4.1. Pyramids, Mastabae, Shaft Graves (Egypt)

### Pyramid No. XXVIII

South of the relics of Ninserre's valley temple (see Fig. 116, sector E) over the essentially flat field of the desert margin there arises a conspicuous hillock, about 10 m high. It is roughly L-shaped, one arm being oriented in the S - N direction, the other from the S to the W. The length of the two arms is practically the same, something more than 100 m. Lepsius (1849-1859) considered this hillock to be the relic of a pyramid and gave it the number XXVIII. In 1907, the object drew the attention of an archaeological expedition DOG (Borchardt 1907). More detailed information about the investigation made at that time was rendered by Borchardt (1910). From his report it follows that several probes were made into the object at its E and S slope and on the top.

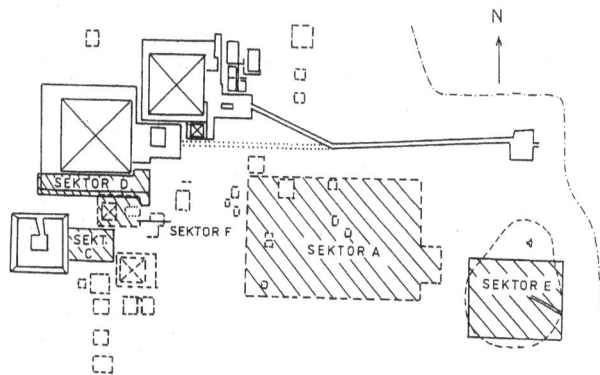

*Fig. 116 Abusir, Egypt. Scheme of necropolis with marked geophysically measured regions.*

The maximum depth of the probes was about 6 m. Their deepening was, however, prevented by ground water. Below a layer of sand about 1 m thick loam was struck which L. Borchardt considered artificially deposited. At the lowest level pottery fragments were found which, in the author's opinion, originated from the end of the Old or Middle Empire. On the whole L. Borchardt declared the object to be a relic of a Middle Empire pyramid and expressed a forecast that the pyramid was linked up with the relics of brick and perhaps limestone buildings stretching from it toward the southeast. Borchardt's idea was also considered probable by V. Maragioglio and C. Rinaldi (1970: 176-184) who, in the whole plan of the Abussir burial ground even marked a square ground plan of an object with the length of a side of 100 m (Hašek - Verner 1981: 306-316; Hašek - Obr - Přichystal - Verner 1986: 149-187).

During the excavation period of 1978-1979 the object became a place of geomagnetometric measurement in the network of 2.5 x 5 m. But the results of the works (Fig. 117) did not confirm Borchardt's belief. It was proved that with great probability it was not the ruins of a brick building, but a natural morphological anomaly in the field relief. Only at the northern and the western slopes of the hillock it is possible to assume small brick tombs. The values of the anomalies ΔT in those places vary up to + 10 nT. In the other sectors of the investigated area the magnetic field has a relatively quiet course. This conclusion is confirmed on the one hand by earlier probes made by L. Borchardt, on the other hand also an intense occurrence of pottery in places of the local anomalies of the magnetic field.

### The Abussir Burial Ground

Geophysical methods employed in the eastern and western sectors of the Czechoslovak licence territory near Abussir were to find the overall extent of the two parts of the burial ground, to verify the position of the individual brick and stone tombs and also of other objects of interest.

The eastern sector (see Fig. 116, sector A) includes an extensive burial ground, the so-called "field of mastabae", with tombs of magnates, high officials and lower members of the royal family. In that space the tomb of Princess Chekeretnebtey was found, which is the highest situated structure in the measured territory of the dimensions of 240 x 110 m, situated on a small morphological ridge with the axis in the W - E direction.

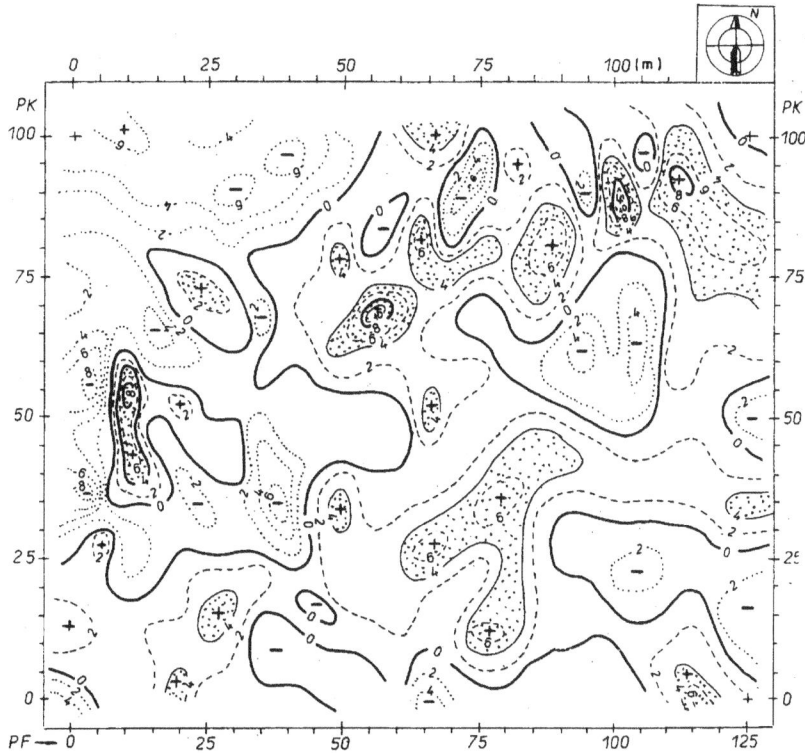

*Fig. 117. Abusir, Egypt. Map of ΔT isanomales in the region of hypothetical pyramid.*

The above building was constructed of small limestone blocks and of dried brick (æ = 0.82 . $10^{-3}$ u.SI). It has a rectangular ground plan and its long axis is oriented exactly in the N - S direction. It consists of an overground part and an underground one. In the overground part there are three chapels with unfinished mastabae and steles in the shape of false doors and a room with wooden statues. In the underground part, accessible by vertical shafts, is the funeral chamber with a limestone sarcophagus.

In its close proximity mastabae of further important functionaries were discovered, with brick walls (æ = 1.3 . $10^{-3}$ u. SI) and steles found in the shape of false doors, wooden statues, sarcophagi, etc. (Verner 1980: 243-268). The building complex of the centralized mortuary rite consists of a court with circular sacrificial tables, small basins and three vaulted brick spaces (æ = 0.53 . $10^{-3}$ u. SI) (Hašek - Verner - Obr 1983: 187-199; Hašek - Obr - Verner 1988: 5-47).

From the map of isanomales ΔT which links up directly with the above group of tombs (Fig. 118) a number of both positive and negative anomalies was separated which, after correlation with the curves of apparent specific resistances, can be interpreted in the following way:

a) positive anomalies ΔT in the form of narrow and elongated zones locate the positions of brick walls

b) areally extensive positive anomalies ΔT can be due to a large accumulation of pottery, such as a filled hollow space, a part of the brick vault of tomb roofs, the brick floor, etc.

c) approximately isometric negative anomalies ΔT are formed by stone tombs (limestone inner paving, ceilings, etc.), and/or the sand filling of brick objects.

By the combination of the results of resistance profiling and magnetometry it was found that increased values of apparent specific resistances and the negative anomaly ΔT locate the positions of stone limestone structures and, on the other hand, the reduction of resistances and positive anomalies ΔT the course of brick masonry (see Hašek - Obr - Verner 1988: 5-47).

From the overall distribution of the individual interpreted structures (see Fig. 118) the studied space can be divided into three sectors:

1. relatively much conspicuous is the wider surroundings of the complex of mastabae of Princess Khekeretnebtey, where also three largest tombs are

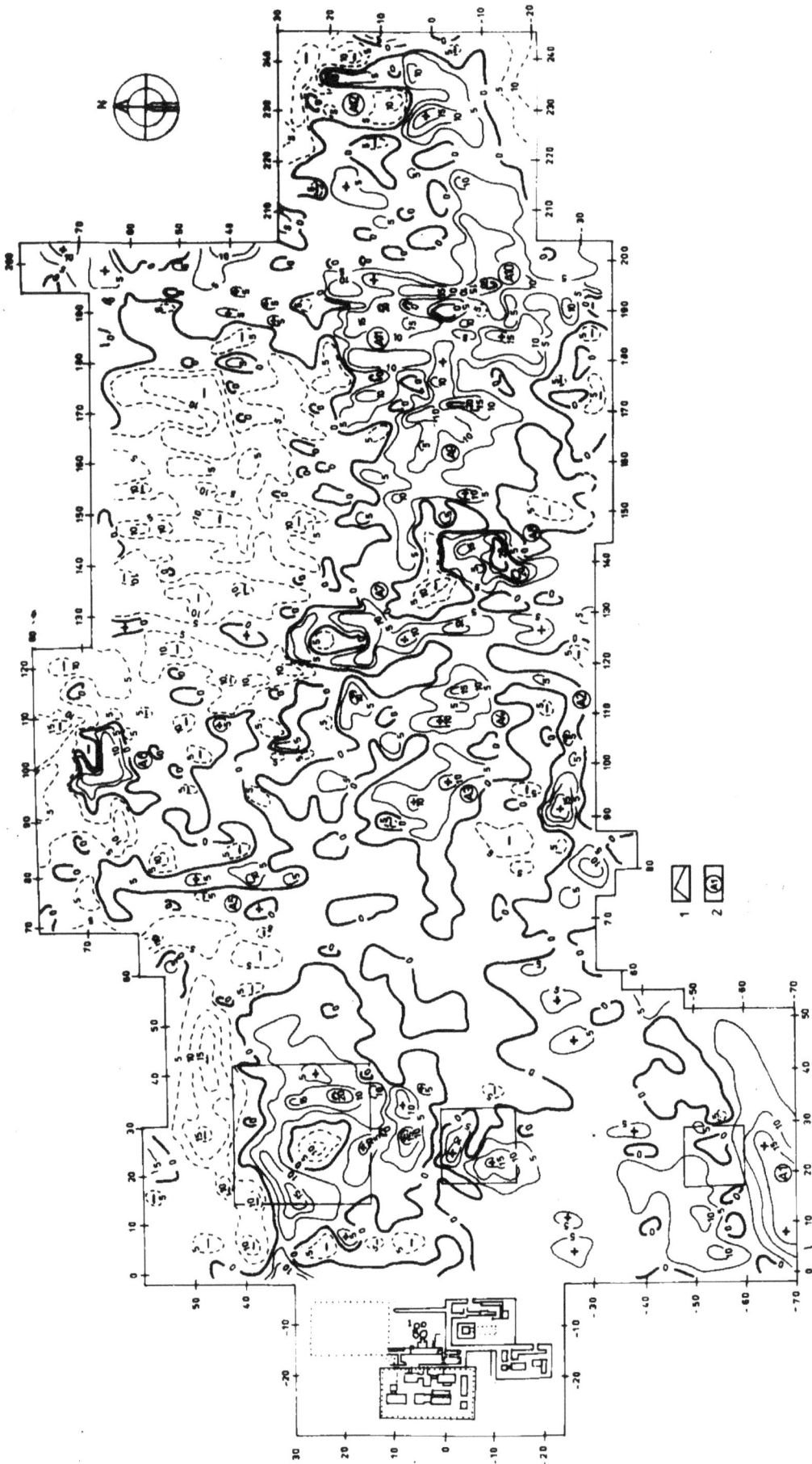

*Fig. 118. Abusir, Egypt. The Mastaba Field. Map of ΔT isanomales and position of tombs.*

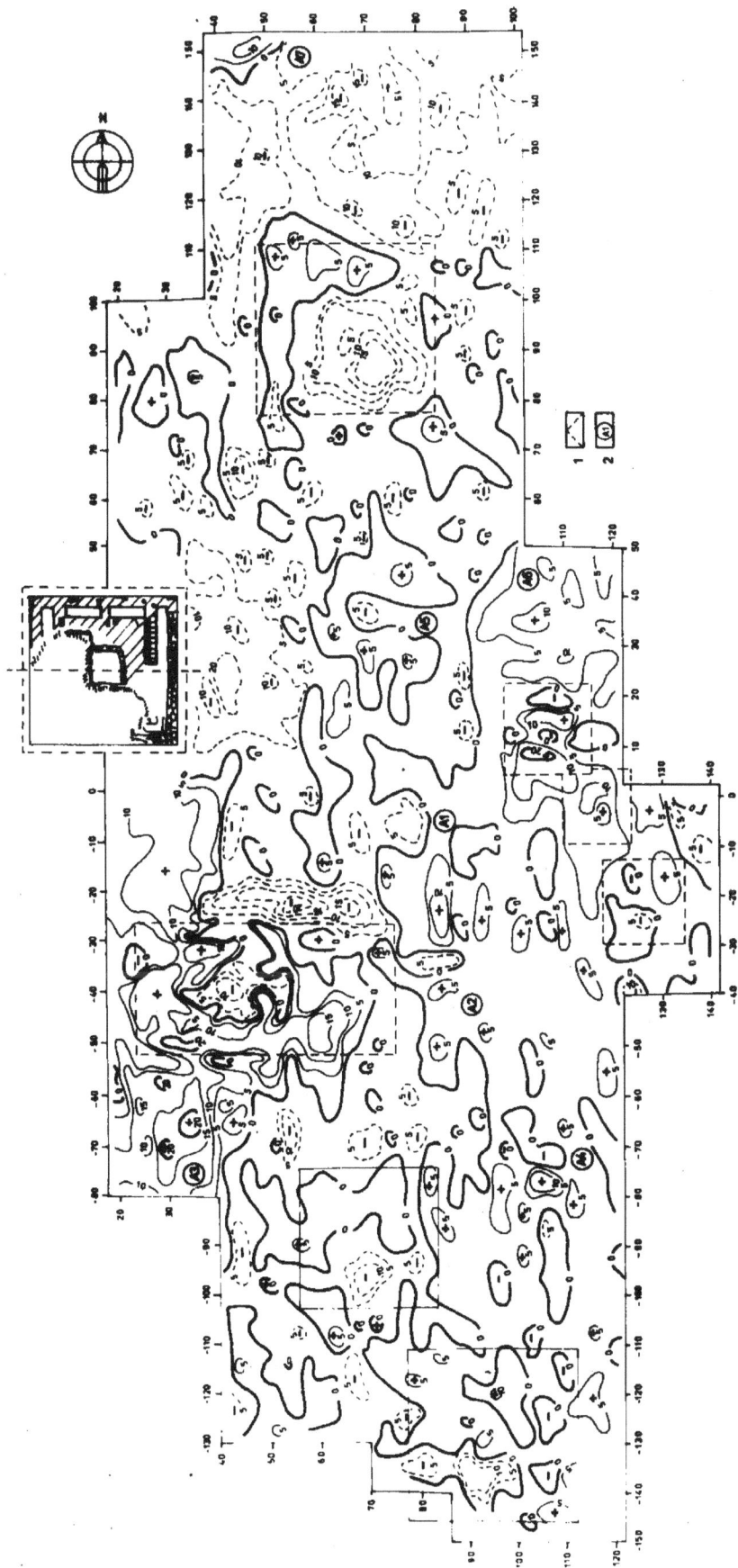

***Fig. 119***. *Abusir, Egypt. Map of ΔT isanomales of the western part of necropolis.*

found, visible also in the relief of the field of the whole measured area.

2. in the central part, on the other hand, individual and minor structures can be assumed, with dimensions up to about 10 x 15 m

3. in the eastern sector a large occurrence of tombs can again be interpreted as well as that of further objects.

The separated structures, excepting the western part of the studied territory, are concentrated prevailingly into the southern part of the morphological ridge. From the results of the processing of the two geophysical methods it can be assumed that in that territory of interest mostly brick tombs were built which, in the central part, can be built of stone blocks - ceiling slabs, or that these blocks constitute their linings and crib walls. It is not uninteresting to see that between the main and highest situated group of mastabae in the surroundings of the Khekeretnebtey tomb and the central part situated already on the slope of the morphological ridge there is a relatively large distance. It cannot be excluded that these two groups will be separated by an enclosure wall, and similar conditions can also be interpreted between the central and the eastern parts of the territory.

A relatively complex character of the field was then found at the easternmost side of the measured territory. A large representation of brick structures can be assumed there, and according to the accumulation of pottery in that space an assumption can be expressed about the possibility of the existence of objects of settlement character (Verner - Hašek 1989: 67-72).

The western sector (Fig. 116, sector B) is situated at the southernmost sector of the licence territory. Originally it was supposed that there would be a small independent burial ground of the same time period as the others in that space. But the burial originates from the Late Egyptian Period and the early Greek Period (Verner 1982). It is quite unknown and more than 2000 years younger. Unique tombs of limestone blocks were found there built by the technology of shaft graves. In their investigation it was found that it was not a single shaft, but a whole complex of them. The central shaft has the ground plan of 5 x 5 m and it is surrounded by a system of peripheral shafts 6 x 2 m (see Fig. 119). All those shafts together form a sophisticated safety system. They are completely filled with fine sand and interconnected. Their hitherto established depth is 15 m (Hašek - Obr - Přichystal - Verner 1986: 149-187). By the archaeological investigation of the Czechoslovak Egyptological Institute, Charles University, it was found that the tomb belonged to Udzahorresnet, one of the most important and at the same time most contradictory personalities of Egypt of the end of the Sai period and the beginning of the 1st Persian supremacy, i.e. the latter half of the 6th century B.C. (Verner 1994: 201).

The investigation area of the dimensions of about 300 x 100 m (Fig. 119) is situated at the eastern to southeastern side of this "shaft superstructure", being located in a space where in the field relief four major objects are apparent, dimensions about 40 x 40 m, and three within about 20 x 20 m. Unlike the eastern sector, where there were mostly structures of dried brick, the chief mastabae in this space are mostly built

of limestone blocks, only the extensive structure to the SE of the shaft grave consists of a combination of brick and stone. The central part of the tomb is of limestone blocks. According to the data of geoelectric measurement also the position of the main shaft cannot be excluded at PF 44 of picket -42 and partial shafts and/or stone walls at PF 42 of pickets -28, -50, -64 and at PF 44 of picket -28, which would confirm the assumption of the possibility of the occurrence of another shaft grave in that space (Hašek-Obr-Verner 1988: 5-47).

Another two structure to the SE of the Shaft grave and one on the northeastern side were built mostly of stone (Fig. 119). Only the circumferential masonry is supposed to be possibly of brick material. From the obtained results a considerable disturbance of the above objects can be assumed.

Foundations of smaller and in the magnetic field less conspicuous objects can be separated at the eastern margin of the measured area.

The course of the circumferential brick wall of the burial ground can be assumed in the northern part of the investigated territory, between PF 40 to 54, in the surroundings of picket 150 m (Hašek-Verner-Obr 1983: 187-199).

### 4.5.4.2. Crypts

*Brno, district Brno, St. Thomas' Church*

The task of measuring by the geophysical radiolocation method carried out in 1984 by the workers of the n.e. Geoindustria Prague (Bílý in Hašek et al. 1985: 22-26) was to verify the extent of the crypts under the triumphal arch for the later effective way of their opening. The objective of the works was to investigate the tomb of the Moravian Margrave Jošt (+1411) and to find the position of the today unknown tomb of his father, brother of the Emperor and King Charles IV, margrave Jan Jindřich (+1375) (Hašek - Měřínský 1987: 102-140).

In Fig. 120 an interpreted position of hollow spaces below the church floor is plotted. From the results obtained the character of the filling of the located cavities cannot be stated unambiguously.

Unambiguously verified by measurement were tombs to the left and to the right of the main altar (see Fig. 120) covered by tombstones with inscriptions. Even though both the beginning and the end of the profile was on them, in the two objects diffracted waves are distinct at the contact of the floor and the wall. A more complicated situation is under a big slab of red limestone, where there are remains of Margrave Jošt (thickness about 0.5 m) and in its immediate surroundings in front of the main altar. Under this slab two types of signals of reflected electromagnetic waves were found. On the part of the record belonging to the sectors in the close proximity of the slab there are intense multiple reflections which can be due to a free hollow space of the height comparable to the wavelength (1 m) or due to the occurrence of metal objects near the surface (Fe-rods) and/or the combination of those phenomena. From the processing of the records, in further sectors partly collapsed cavities are

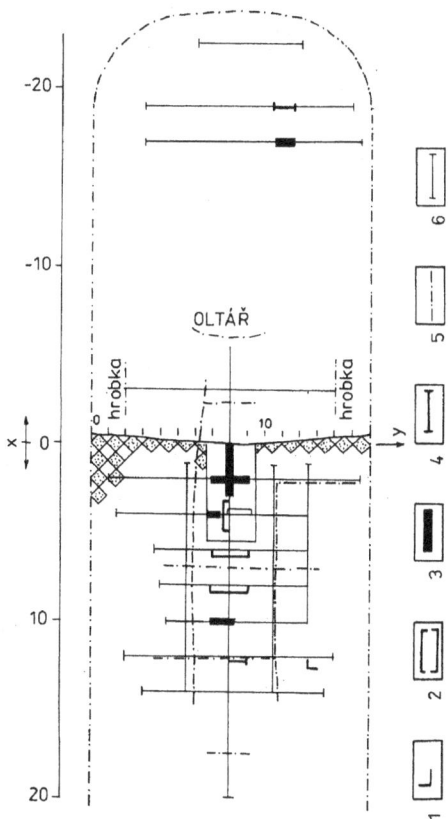

*Fig. 120. Brno. GPR situation and interpretation. St. Tomáš church. 1- edge, 2- small (possibly filled) cavity, 3- multiple reflection, 4- reflection interface, 5- energy networks, 6- geophysical profile.*

assumed in which the free space is of irregular shape (absence of diffracted waves).

In the space behind the altar a record of good quality of a multiple reflection at profile - 17 was obtained. Besides, the reflections on the neighbouring profile can be explained as lateral ones from an anomalous body resting directly under the profile. With respect to a considerable intensity of the reflected signals this anomaly can be explained as a sign of a cavity, the same as that of a metal object at the depth of 0.5 m below the present floor. Excavation works have not been started there so far, even despite the fact that some data obtained by measurement correspond to the description of the adaptation of the tomb and deposition of the remains ofMargrave Jošt in the Baroque adaptation of the church in the first half of the 18th century, as stated in the monastery chronicle (cf. Hašek - Měřínský 1991: 141-142).

*Křtiny, district Blansko*

In the Baroque pilgrimage temple of St. Mary in Křtiny, the geophysical investigation by microgravimetry and the DEMP method was directed on finding the possibilities of using the above methods for the location of tombs of different sizes and graves under conditions of sacral buildings.

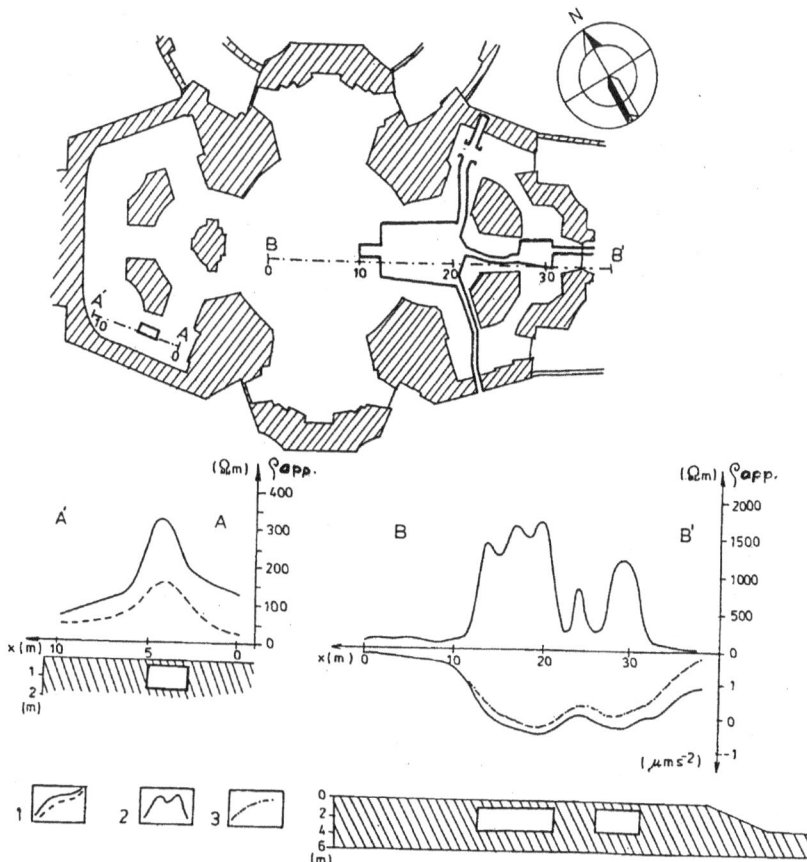

*Fig. 121. 1- curves $\varsigma_{app}$ for h = 1,5 m, respectively h = 4-5 m, 2- $\Delta g_{nam}$ curve, 3- $\Delta g_{red}$ curve*

On the longitudinal profile led in the axis of the church (Fig. 121) with the step of measurement of 1 m, from the results of gravity measurements (Bednář - Novotný - Švancara 1980: 23; Hašek - Měřínský 1991: 143-144) two local negative anomalies g were found accompanied by increased values of specific resistances of the DEMP method. The first of them is over a known crypt, the second, SE of the first one, indicated - after the investigation - a charnel-house - ossarium (Šenkyřík 1992: 8; Hašek - Unger 1994: 33).

A detailed measurement (the step of 0.5 m) by the DEMP method (h = 1.5 m, h = 3 - 5 m) carried out in the SE part of the church, in places of the earlier exposed shallow grave of the abbot K.J. Matuška, indicated its position by a belt of increased resistances also on the whole unambiguously.

## Znojmo, district Znojmo

The task of archaeogeophysical prospection in the investigation of the Romanesque crypt (the turn of the 12th and the 13th centuries) of St. Wenceslas' Premonstratian (Premonstrate) monastery at Znojmo-Louka was to locate the positions of assumed inhomogeneities under the floor. By measurement with the apparatus of DEMP of different depth ranges (h = 1.5 m and h = 3 - 5 m) and with the ground radar, in its interior, approximately in the axis of the building, despite some disturbing effects (cables, lateral effects of masonry) altogether three anomalous regions A, B, C were found (Fig. 122), accompanied by the impairment of the brick floor. The dimensions of the separated inhomogeneities vary from 1.5 x 2.5 m up to 5 x 2.5 m. The main axes are oriented in the E - W direction (Hašek et al 1995a).

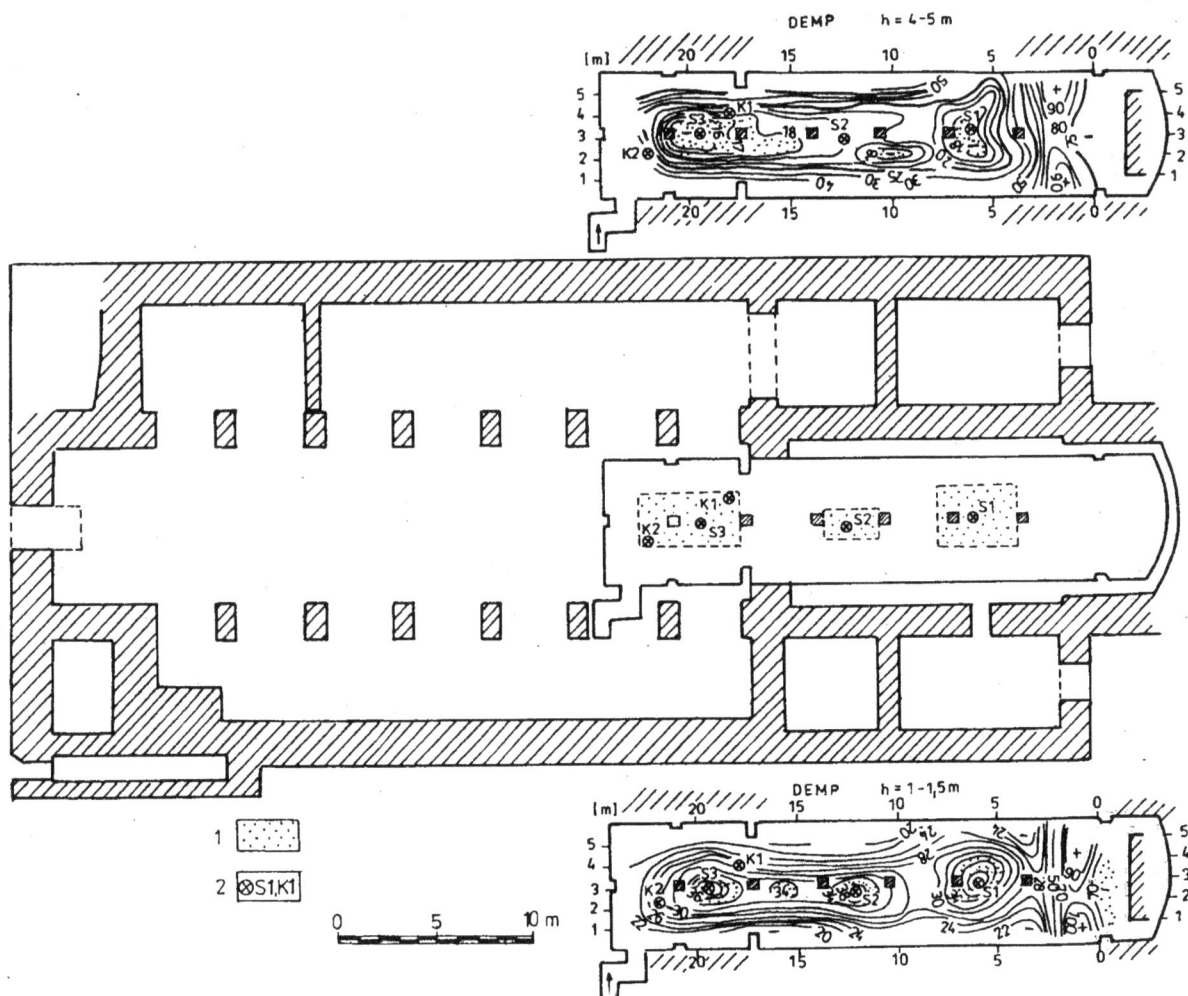

**Fig. 122.** 1- local inhomogenity, 2- verification drill holes

From the results of the verifying pedological probes (5 stabs) of all three anomalies (see Fig. 122) there can be assumed a combination of effects of earlier recessed objects at greater depths under the floor of the crypt, partly filled with building and other material, accompanied at smaller depths with half-filled cavities (a grave, a tomb) and/or covered with a stone slab. A possible combination of large butts or a belt for grounding the central pillars of the crypt and recessed objects - cavities cannot be excluded.

# 4.6. Prospection of Mining Centres of Mineral Raw Materials

Geophysical works carried out for the purposes of prospection of exploitation centres of mineral raw materials were directed above all on finding the remnants of medieval to modern surface and deep-mined exploitation of mineral raw material, above all gold, ferrous and non-ferrous metals, coal, etc. in connection with the revision and engineering-geological investigation of the stations of interest in Moravia. Neolithic and Eneolithic loam-pits for extracting loam for plastering the walls of overground houses, etc. are not mentioned here. For solving the required tasks, e.g. the location of cavities (galleries, shafts), altered zones, etc. above all geoelectric methods were employed, completed by magnetometry. The use of microgravimetry for the above purposes are described e..g. by Odstrčil in Hašek - Měřínský 1991: 45-47).

## 4.6.1. Surface Traces After Mining

*Hory, district Třebíč*

In the surroundings of the village of Hory numerous documents about the extraction of mineral raw materials have been preserved (dumps, remains of well pits, etc.) dating back to the period of culminating Middle Ages. There are traces of mining and panning works for Au of the 13th to 15th centuries (Merta 1984; Měřínský 1984).

Geophysical works carried out in 1987 for purposes of the revision investigation of the n.e. Geoindustria Brno at an area of interest of 400 x 450 m by the VLW method in the resistance and the inductive versions and by magnetometry at the step of measurement of 5 m were to find the position of altered zones in the varied series of Moldanubicum (biotitic gneisses, amphibolites, quartzites, crystalline limestones, etc.) to which occurrences of Au can be bound (Hašek-Měřínský 1991: 144-145).

By geophysical measurement using the VLW methods in both versions (Fig. 123) a number of weakened zones was found which in places follow the dumps. Anomalies of the magnetic field in the eastern and southeastern parts of the studied area (+ 20 nT) are probably due to positions of quartzites containing pyrrhotine (æ = 1.17 $10^{-3}$ u.SI)
The verifying groove made by a motor dredger approximately through the centre of the studied area demonstrated the existence of all interpreted altered zones of large thicknesses and further the position of mineralized quartzites. From the viewpoint of Au occurrence these deformations are quite negative in their near-surface part (verbal communication - Dr. I. Mrázek), which outlines the possibility of finishing the medieval mining works in that space due to the exhaustion of the deposit. The occurrence of Au bound to the separated zones of alterations was found by boring only in greater depths (Hašek - Měřínský 1989: 103-151).

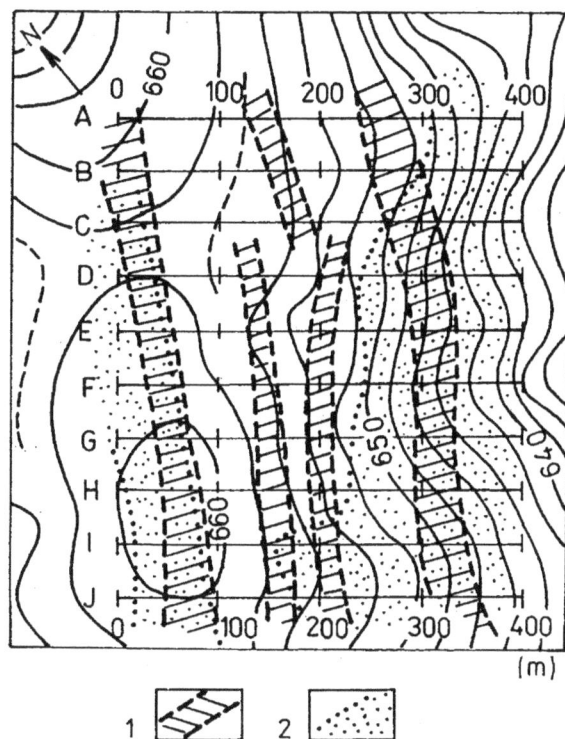

**Fig. 123.** *Hory, district Třebíč. Correlation scheme of geophysical results. 1- interpreted course of conductive zones - alteration zones, 2- position of magnetic rocks.*

## 4.6.2. Subsurface Relics of Mining Works of the Modern Period

*Hodonín - Mikulčice, district Hodonín*

Near Mikulčice the Jihomoravské lignitové doly, k.p. (South Moravian Lignite Mines) built a new mine Hodonín I. In the building and excavation works it was found that in the space of the building site there occurred unblown, partly blown and founded galleries from the extraction of the lignite bed whose age was estimated to be about 100 years. Thus an imminent danger occurred that due to the subsidence of overlying beds to the extracted spaces the individual building objects might be damaged. The assumed depths of the cavities were 4 to 8 m, their dimensions 1 x 1.5 to 1.5 x 2 m.

Geophysical works performed there by geoelectric methods (three-electrode gradient arrangement - AMN, the resistance version VLW) at the step of 1 m and shallow refraction seismicity in 1980 was to locate the above spaces in places of intensely disturbed by building and other activity (Hašek - Pantl - Zemčíková 1980; Hašek - Ilčík 1980).

In the place of geophysical works there overlying as well as soft underlying beds (Pliocene, Panon s.s. - upper Panon) consist of sand with different amounts of admixtures of clays, sometimes also gravel. In places these beds pass into sandy clay (Hašek et al. 1981: 58-59). The values of the apparent specific resistance measured by a potential probe (Rap) varied from 10 to 30 ohmm (lower values for beds with a higher content of clays, higher ones for sandy beds). This interval also includes the values of Rap of the bed (20 - 30 ohmm). With respect to the resistance of the drilling fluid and the diameter of the borehole it can be assumed that the actual specific resistance can be a double of the above specific resistance. On the whole it is possible to state that the values of the resistances in the environment studied are low and that they differ very little from each other. The overlying beds, the bed and the underlying beds can thus be considered to be an approximately homogeneous layer. The cavities not filled with water can be imagined to be inhomogeneities in that layer with increased specific resistance.

From the summary processing of the results of geoelectric measurement in the form of a correlation diagram from the space of the building site (Fig. 124) and in comparison with the boring data - borehole 33 - 80, 34 - 80 and 8 - 80, situated on the basis of the geophysical interpretation on profile A (Fig. 125) it followed that the separated zones of increased specific resistances locate prevailingly the positions of galleries in the lignite bed whose roof in the dredged out area varies at the depths of 6 to 7 m, at the present field surface at 9 - 10 m.

From Fig. 124 it is evident that both the exposed area (profiles A, A1, B) and the space in their proximity (profiles C, D) was affected by a relatively intense mining activity, the interpreted courses of galleries being mostly in the N - S direction, with branches into the E - W direction. The wider zones of increased apparent specific resistances can be a sign of two parallel galleries, crossings of corridors, etc. (Hašek - Měřínský 1991: 148-150).

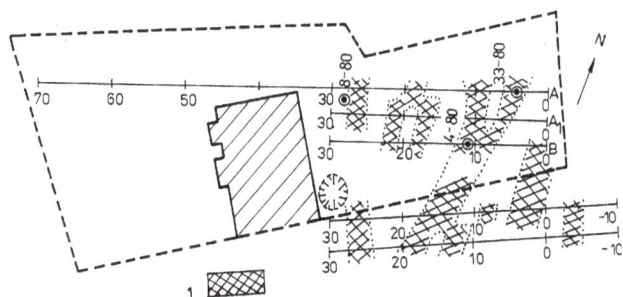

**Fig. 124.** *Hodonín - Mikulčice, district Hodonín. Correlation scheme of geophysical results. 1- interpreted course of undermined areas*

The performed experimental seismic measurements have demonstrated that under the given seismological conditions it cannot be used for looking for corridors of such a small cross-section.

*Vratíkov near Boskovice, district Blansko*

In the broad surroundings of the community of Vratíkov, the space of the Devonian limestone massif (the development of the Moravian Karst) relics of medieval to modern mining activity were found (dumps, etc.). A gallery following the vein of magnetite-hematite ores in the NNE - SSW direction

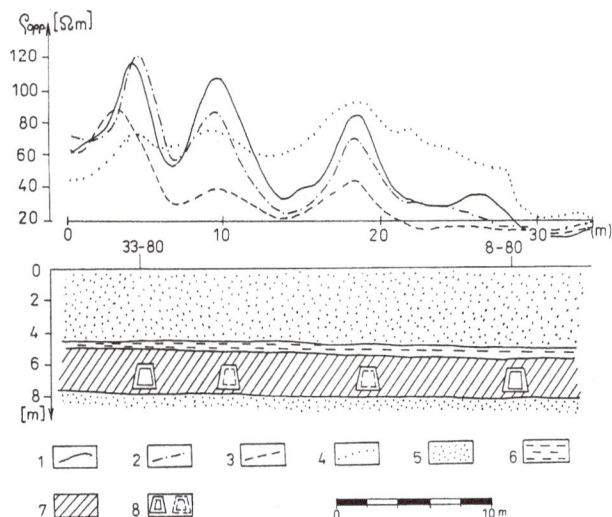

**Fig. 125.** *Hodonín - Mikulèice, district Hodonín. Profile A - geological and geophysical profile. 1- $\rho_a$ curve obtained from A5M2N electrode setup (m), 2- $\rho_a$ curve obtained from A8M2N electrode setup (m), 3- $\rho_a$ curve obtained from A13, 5M3N electrode setup (m), 4- $\rho_{VDV}^H$ curve, 5- dusty sand, clay sand, 6- coal clay, 7- lignite, 8- found and interpreted galleries.*

has been preserved, with several falls of roof. The depth of its roof below the field surface is about 3 to 5 m, the height 2m, the width 1.5 - 2 m (Hašek-Měřínský 1991: 145-148).

The objective of geophysical works performed by geoelectricmethods (AMN-B, MAN) was to follow the extent of undermining in connection with a detailed engineering-geological investigation of the transfer of the road near the water work of Boskovice. From the results of the measurement it followed that the position of the known gallery can be followed only to the assumed fall of roof (Fig. 126). Its further continuation was not demonstrated in the above direction. Another zone of increased resistances starts from the probably fallen entrance to the gallery at PF 2 PK 0 m and reaches to PF 3 with approximately the same picket, where a possible connection can be expected with the above zone of increased resistances, which indicates a more complex course of this system (fall of roof, break in the trace) than has been assumed before (Hašek - Unger 1994: 33).

A relatively conspicuous resistance anomaly can also be separated at PF 3 PK 14 m which continues in the northern direction up to PF 5. In this case the sign of a cavity cannot be excluded, possibly linking up with the course of the above gallery in greater depth, which is witnessed by inbreaks situated approximately into the axis of this interpreted nonconductive zone. In further interpreted noncoductive zones only effects of certain lithological changes in the massif can probably be assumed (karsting and cracking of carbonates) as an indication of further near-surface cavities.

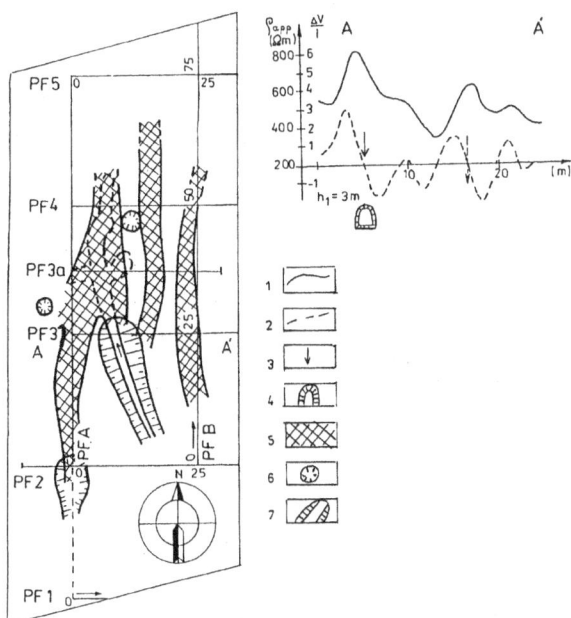

*Fig. 126. Vratíkov near Boskovice, district Blansko. Correlation scheme of geophysical results and comparative profile at the mouth of a found gallery. 1- $\rho_a$ curve obtained from A5M2N electrode setup (m), 2- M2A2N curve (m), 3- interpreted nonconductor, 4- gallery, 5- zones of increased resistance, 6- filled gallery, 7- course of a found gallery.*

## 4.7. Prospection of Production Objects

The task of archaeogeophysical prospection of production objects at the stations of interest, particularly in the region of the Boskovická brázda (the Boskovice Furrow) and the Moravian Karst was to determine the position of furnaces and/or other objects connected with smelting, whose existence is supported by the occurrence of Fe-slag, fragments of nozzles, daubs of furnaces, etc. Extensive tips of waste slags and burned layers of furnace batteries have a marked differential magnetization towards the surrounding environment, and from that point of view magnetometry asserts itself as the optimum method.

### 4.7.1. Metallurgical Equipment

*Olomučany, district Blansko*

At the station "U srnce" (the Red Deer) in the cadaster of Olomučany, at the area of about 27 x 22 m measured by magnetometry (Fig. 127) relics of a 9th century A.D. workshop were exposed (the Middle Ringwall period, the Great Moravian period). The workshop produced iron and it was a part of a mining-metallurgical area extending in the central part of the Moravian Karst. The area existed from the 8th to the 11th centuries A.D. and 15 metallurgical workshops with different types of furnaces have been investigated so far. Each of those metallurgical workshops operated for only a certain time - until the trees near the workshop were all felled and processed into charcoal which was the only fuel for then used furnaces.

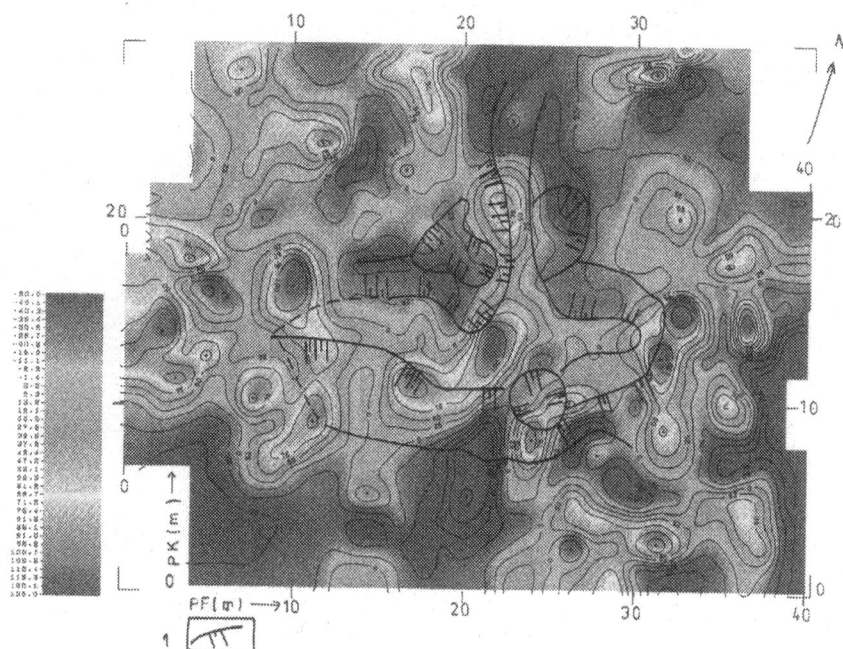

*Fig. 127. Olomučany, district Blansko. Map of ΔT isanomales and situation of archaeological research. 1- position of excavated furnaces.*

In the place of one of the measured anomalies ΔT (Fig. 127) the relics of two furnaces were found at its edge. That is the so-called built-in type with a thin breast about 0.6 - 0.7 m high and the inner diameter of 0.5 m in the hearth. Below the

furnaces waste tips with slag were found, fragments of furnace walls, nozzles, pottery, etc. The tips reached the maximum height of today 0.7 m. The waste tip below the furnaces creates intense magnetic anomalies due evidently to the accumulation of iron slags which in the then used direct production of iron still contained a considerable amount of ferromagnetic minerals.

*Sudice near Boskovice, district Blansko*

In the cadaster of the community of Sudice, in the tract "U občin" numerous traces of metallurgical activity were found dating back to the La Tene and the Late Roman periods (Hašek - Měřínský 1989b). These relics of metallurgical activity were, however, scattered due to field works on an area larger than 2 ha, so that finding their sources by classical archaeological methods appeared to be rather difficult and time consuming.

For those reasons, for the indication of metallurgical installations magnetometry was applied in 1975 (Hašek - Mayer et al. 1976: 6-14) and in 1976 (Hašek - Ludikovský et al. 1977: 25-30); with respect to different magnetic properties of the object of interest (æ = 3.8 - 22.6 x $10^{-3}$ u.SI) and the surrounding environment - permocarbon of the Boskovice Furrow, opsoil (æ = 0.13 - 0.63 x $10^{-3}$ u.SI) - magnetometry can be considered to be the most suitable method for obtaining optimum data for investigation in the field, carried out by the AI CSAS in Brno and by the District Museum of Local History at Blansko. At three measured areas of 50 x 26 m (A), 20 x 10 m (B), 27 x 15 m (C) in the network of measuring points of 1 x 1 m (Fig. 128) altogether 5 approximately isometric anomalies ΔT were found which were recommended for further archaeological checking. The result of the computer processing of area A by means of bicubic splins, together with the exposed objects, is given in Fig. 129 (Hašek-Horák-Obr 1979). The spatial representation of anomalies ΔT is in Fig. 130.

From the overall character of the magnetic field (Figs. 129 and 130) two intense anomalies ΔT (+ 50 nT, + 8 nT) can be separated. In the investigation season of 1975, in the place of the former, after removing the topsoil by a bull-dozer and cleaning off the surface a concentration of 72 metallurgical furnaces of the Late Roman period (the 3rd to 4th centuries A.D.) on an area of only 6 x 6.5 m was found (see Fig. 129), oriented in the NW - SE direction (Ludikovský - Souchopová - Hašek 1977). In an intact state only their hearths were found filled with slag, remnants of charcoal, etc. The diameter of the individual furnaces of 0.5 m varied within the limits of - 0.08 to + 0.05 m. Towards the bottom the hearth cavities extended conically and they passed continuously into dish-like smoothed furnace bottoms, in some cases with traces of daub of the walls with clay (Burdová - Hašek - Ludikovský - Obr - Souchopová 1977).

Linking up with the results of the field excavations in 1975, in 1976 further works were concentrated on the one hand on the latter anomaly ΔT (+ 8 nT) from area A (Fig. 129) and, on the other hand, on area B. The former yielded a large pit serving as an experimental device for solidification, filled with ash-loamy filling containing a large amount of charcoal, fragments of walls of metallurgical furnaces, further a battery of 17 hearths of furnaces analogous to the above ones (Ludikovský - Souchopová - Hašek 1977).

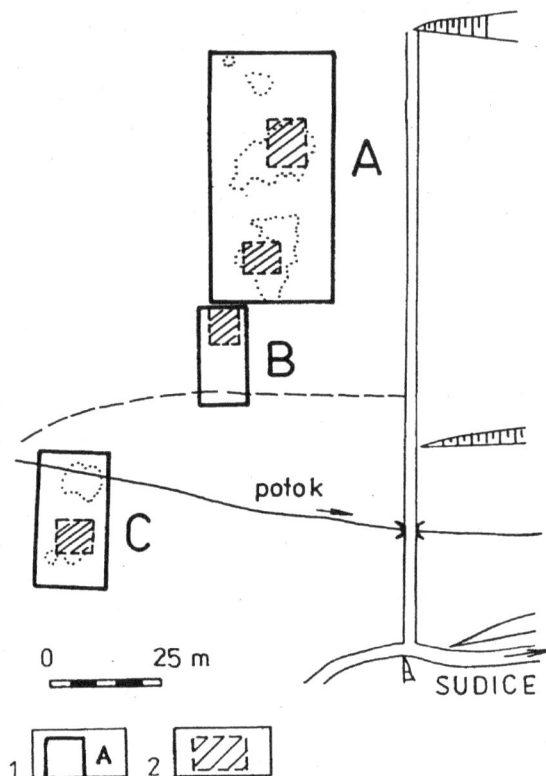

*Fig. 128 Sudice near Boskovice, district Blansko. Situation of areas under geophysical measurements and archaeological research.1- measured areas, 2- situation of excavations.*

The excavation at the more southern anomaly ΔT (+ 53 nT) of area C (Fig. 131) and the subsequent archaeological investigation in 1977 (Ludikovský-Souchopová 1978) found relics of 47 hearths of metallurgical furnaces. In a considerable part of them their circumferential jackets were preserved, in isolated cases up to the lower level of the smelted iron.

The last measured anomaly ΔT (+ 60 nT) in that space is situated north of the investigated area in the lowest part of the tract in the water-logged part of the field. Its investigation would be possible only after drainage.

The number of hitherto excavated furnaces exceeded 136 installations, and the find place, according to the results of geophysical works, cannot be considered to be exhausted. Remarkable is above all the uniqueness of the conglomeration in the Moravian environment (Hašek - Měřínský 1990: 152-157).

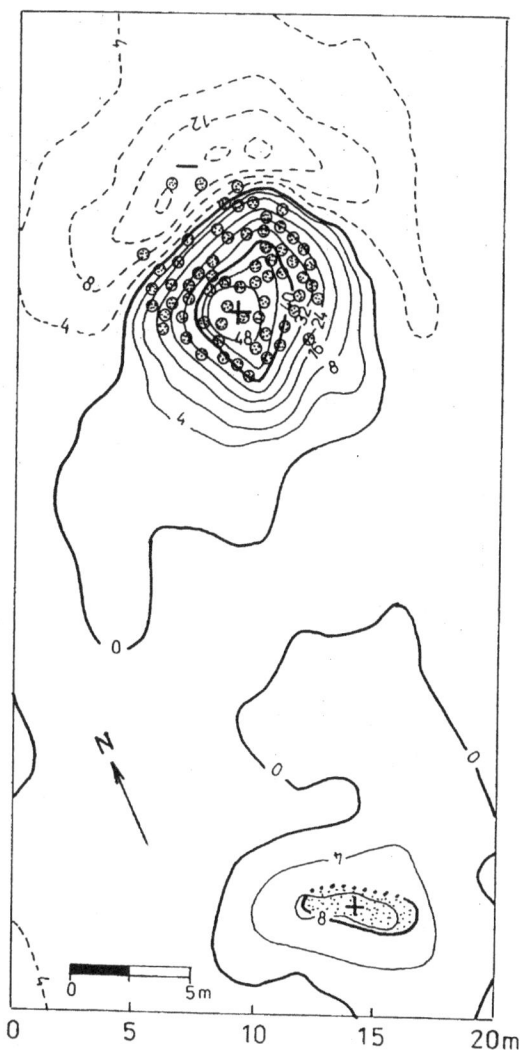

**Fig. 129.** Sudice near Boskovice, district Blansko. Map of ΔT isanomales with marked excavated archaeological objects - region A.

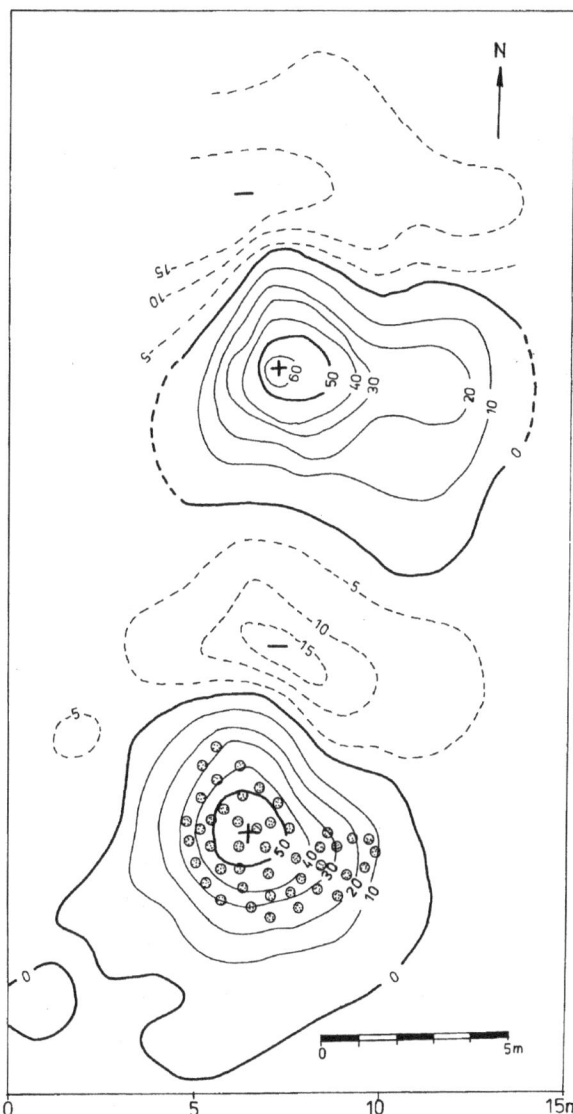

**Fig. 130.** Sudice near Boskovice, district Blansko. Axonometric depiction of ΔT isanomales in region A.

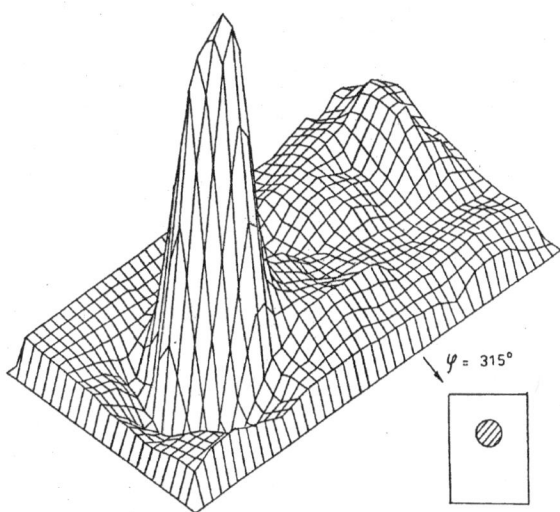

**Fig. 131.** Sudice near Boskovice, district Blansko. Map of ΔT isanomales with marked excavated archaeological objects - region C.

### 4.7.2. Other Production Objects

*Drválovice near Boskovice, district Blansko*

Some new information from the surface investigation of stations in the area of Mal Han indicate that the occurrence of metallurgical appliances is not limited only to the cadaster of the community of Sudice near Boskovice, but that it takes up a substantially wider territory. On the basis of a preliminary investigation in the field in the surroundings of the community of Drvalovice-Vanovice near Boskovice, in the tract "Přední Smolínka" a large concentration of slag material was found (Hašek-Měřínský 1991: 158-160).

In 1977, at the area of interest of 20 x 20 m and in the network of 1 x 1 m magnetometric measurement was carried out for the AI CSAS in Brno with the purpose of locating the assumed metallurgical installations (Hašek et al. 1977: 24). The resulting processing (Fig. 132) separated two approximately isometric anomalies ΔT (+ 200 nT, + 50 nT) of interpreted dimensions of disturbing bodies about 5.5 x 3 m and 3 x 3 m, on which an areal excavation was concentrated in 1978.

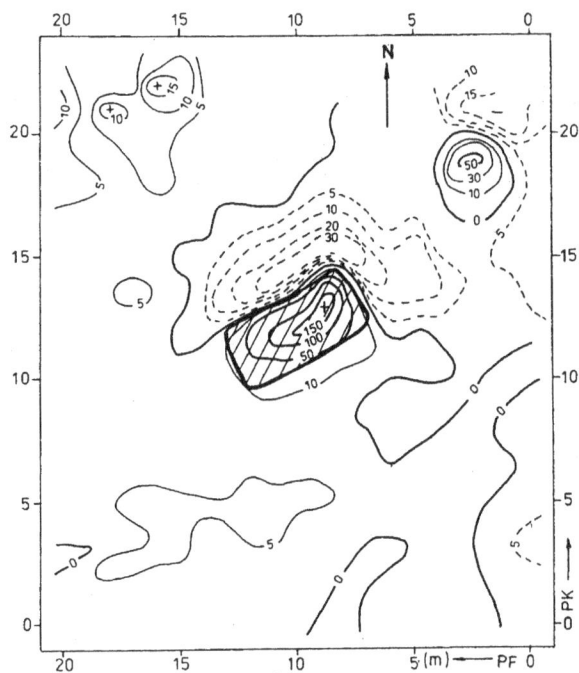

**Fig. 132**. *Drvalovice near Boskovice, district Blansko. Map of ΔT isanomales and position of the excavated object.*

After removing the topsoil layer, by detailed investigation in other places of the intensive anomaly ΔT (+ 200 nT) a metallurgical furnace of rectangular shape with rounded corners was exposed of the dimensions of 1.9 x 4.6 m, oriented longitudinally in the NE - SW direction. The depth of the pit from the interface of the topsoil - sterile environment - reached 0.2 m. The surface of an almost flat bottom, mildly sloping to the southwest, was covered with a continuous layer of charcoal pieces. Their overlying layer consisted of accretions of iron slag whose thickness was as much as 0.3 m and they covered 62 % of the whole area (Ludikovský 1980). In the space of the second anomaly ΔT (+ 50 nT) damaged relics of only one furnace were found. The dating of the found object was not documented by artifacts. On the basis of the external comparison of its shape it is possible to consider its inclusion in the Later Roman period (Ludikovský 1980: 63-64).

# 5. SUMMARY OF THE RESULTS AND OF NEW INFORMATION

This chapter summarizes theoretical as well as practical possibilities of the application of applied geophysics, above all, however, proton magnetometry and different geoelectric methods in solving different tasks of field archaeological prospection and investigation. It contains a set of papers focused on a series of tasks (methodology of works, interpretation, practical examples, etc.) from the above set of issues as well as in a broader sense (assertion in engineering geology, building, etc.).

As documented by the results, in the last decade geophysical prospection has become an inseparable part of the comprehensive archaeological research particularly in Moravia, under conditions of the arid territory of Egypt, etc. In connection with the excavation works or probing they secure the optimum solution of the required tasks within the systematic research and rescue operations from the points of view of technique, time and economy. Practice has, however, shown that these three factors are usually interdependent, sometimes working in opposite directions. Obtaining a great number of data can bring along a disproportionate rise in the price of geophysical works. A great reduction of the time of measurement, such as a smaller area studied on an archaeological structure, a thin network of points, etc. on the other hand causes the reduction of elaboration and detail of the information acquired. From that point of view the choice of the optimum complex of geophysical methods is generally rather complicated and it cannot always be stated exactly for every task to be solved. It depends on the overall geological and/or archaeological situation and the degree of investigation of the station, as well as on the character of the task, etc. Rationally chosen geophysical methods permit to reduce the volume of the often only accidentally selected excavation sites or probes (according to surface collections, situation in the field, etc.), by their purposeful distribution. Probes and excavations are suitable to be situated, according to previous experience, above all to the sector of interest, separated on the basis of the results of geophysical works. The obtained information from field investigations can then be utilized for further works e.g. of a more detailed character and/or for making a new interpretation of the whole measured and hitherto archaeologically not investigated structure. If, however, only one method suits the solution of the required task, it would be uneconomical and time consuming to apply also further disciplines.

Geophysical works connected with the solution of geological conditions and the archaeological situation at the individual stations of interest in the Czech Republic, Egypt, Germany and others, can generally be divided into the mapping of the object of interest (fortifications, settlement, graves, production objects, sacral buildings, cavities, etc.) and the determination of limiting conditions - connection with the surrounding environment, i.e. overlying and underlying beds, etc.. As for the actual works, it is possible to recommend above all highly productive methods, such as magnetometry and different variants of geoelectric methods - SRP, DEMP, Georadar (further geophysical disciplines have, with respect to their productivity, time requirement, etc. a smaller assertion) starting from the information that individual archaeological memorials have different physical properties (particularly magnetic and electrical) which are reflected directly or indirectly in the results of geophysical measurements. In some cases the comparable values of electric parameters can differ by their geological and/or archaeological representation. In those cases ambiguous physical information is obtained and the interpretation will only be probable. Sometimes this probability can be very high and vice versa. Also for those reasons the application of a rational complex of geophysical methods is advisable. The elaborated methodology of measurement and the interpretation of the measured data starts from that assumption.

A general overview of using magnetometric and geoelectric methods suggested for solving different partial tasks under the conditions of the CR, Germany and Egypt,compiled on the basis of experience of long years in the above set of issues is submitted in Tables 6 and 7. These tables include, for the sake of completeness, also potential possibilities of further geophysical disciplines.

Besides the chiefly applied methods for solving the individual archaeological tasks, such as magnetometry and the geoelectric resistance method, DEMP (conductometers), VLW in the resistance version, georadar, etc. can also be recommended for some special special tasks. The practical application of all these above disciplines is conceived in the work both from the point of view of expediency and economy, and in dependence of the limiting factor which is e.g. the size of manpower, etc..

For a more credible mutual correlation of anomalous zones and further also from the viewpoint of the machine processing of the measured data above all a regular network of profiles is suitable. Their lengths and distances must, however, be adapted to the character of the archaeological investigation in the field and to the character of the task.

As is evident from the submitted results, for a credible interpretation of the measured data the finding of geological properties of rocks is important, both in the laboratory method and in situ (parametric measurements, logging works, measurements on the walls of, say, exposed objects, etc.). For the application of the method it is chiefly the determination of magnetic susceptibility, specific resistances, permittivity, etc.

The objectives of the publication were reached gradually. A number of programs permitting the processing of the measured data based on different computer technology currently accessible to both geophysical and archaeological institutions were elaborated. They were oriented on the primary evaluation of the field data in the form of plotting maps of isolines, shadow maps and the areal representation of the measured magnitudes, their axonometric representation, etc., qualitative interpretation (the separation of measured fields, calculation of correlation coefficients, etc.) and quantitative interpretation (modelling, deconvolution) in the processing of profile and areal measurements of the methods employed. The suggested methods of numeric processing accelerate - with a large

**Table 6:** An Overview of Application of Magnetometric and Geoelectric Methods

| Subject of Investigation | Applied Geoelectric Methods | | | | | | | | | |
|---|---|---|---|---|---|---|---|---|---|---|
| | Magnetometry | Direct-curent Geoelectric M. | | Electromagnetic M. | | | Seismic M. | | Microgravimetry | Thermic M. |
| | | RP | VES | DEMP | GPR | Induction Detectors | ≤300Hz | >300Hz | | |
| **Open Setlement** | | | | | | | | | | |
| -stone wall | • | + | • | + | + | | + | | • | • |
| -brick wall | + | + | • | + | + | • | + | | • | • |
| -sunken objects | + | • | | • | | • | | | | • |
| -stake holes | + | • | | • | | • | | | | • |
| -furnaces, fireplaces | + | | | | | • | | | | • |
| **Fortification** | | | | | | | | | | |
| -stone walls | • | + | • | + | + | | + | | + | • |
| -wood-and-loam vallums | • | • | | • | • | | | | • | |
| -bunt vallums | + | • | | • | • | | | | | • |
| -soil vallums | • | + | + | + | • | | • | | • | |
| -moats | + | + | + | + | • | | • | | • | |
| **Burial Grounds** | | | | | | | | | | |
| -flat burial grouns | + | • | | • | | • | | | | |
| -tumulusses | • | + | • | + | • | • | | | | |
| -stone/brick graves | + | + | • | + | + | • | • | • | + | + |
| **Production Object** | | | | | | | | | | |
| -iron furnaces | + | | | | | • | | | | • |
| -pottery furnaces | + | • | | • | | • | | | | • |
| -furnaces and fireplaces | + | | | | | • | | | | • |
| **Other Objects** | | | | | | | | | | |
| -caves | | | | | | | | | | |
| -cellars | | | | | | | | | | |
| -deposits | | | | | | | | | | |
| **Exploatation Centers** | | | | | | | | | | |
| -shafts | • | + | + | + | + | | + | • | + | + |
| -galleries | | + | + | + | + | | • | • | + | + |
| -goldwashing places | • | + | + | + | | | • | | | |

+ - main methods          • - auxiliary methods

number of magnitudes - the possible precisioning of the physico-archaeological model of the studied territory which must be considered a system of abstract disturbing bodies and anomalous effects due to them, approximating the archaeological situation at the station and with a precision essential for modelling expressing in general its structure, its size, shape, physical properties of artifacts and other inhomogeneities, etc. The creation of this model for a comprehensive evaluation of archaeogeophysical prospection in the region of interest can, from hitherto information, be generally divides into the following gradual operations:
1) setting the task for geophysical works,
2) selecting the object of modelling (fortification system, settlement object, grave, etc.),
3) determining (such as according to the analogy of size, shape, physical properties of the modelled structure and the surrounding rock environment, separation of the objects differing sufficiently according to the physical properties,
4) forming the magnetic or physical model and the solution of the direct task for the projected geophysical methods,
5) selecting the optimum complex of geophysical methods and the methodology of measurement.

The precisioning of the physico-archaeological model can be implemented after finishing the field works. This stage can be divided into
a) checking the identity of the starting model and the measured data,
b) forming a more accurate model on the basis of qualitative interpretation,
c) solving reverse tasks for the employed geophysical methods (determination of geometric and physical parameters of anomalous bodies),
d) setting up the resulting physico-archaeological model of the studied structure for the subsequent field investigation.

Note: The first and the second points of model formation must always be implemented in close co-operation with the archaeologist on the basis of a priori information (archives reports, aerial photodocumentation, etc.). In the third phase results of geophysical works carried out at the stations of the same culture etc. are already utilized. Checking of the accuracy of the starting model and the measured data is carried out by comparing the character of the calculated and the measured fields.

**Table 7:** A General Overview of Magnetometry and Geoelectric Methods Applied in Research of Objects From Ancient Egypt

| Object of Investigation | Building Material | Magnetometry | Direct-curent Geoelectric M. | | Electromagnetic M. | | | Seismic M. | | Microgravimetry | Thermic M. |
|---|---|---|---|---|---|---|---|---|---|---|---|
| | | | RP | VES | DEMP | GPR | Induction Detectors | ≤300Hz | >300Hz | | |
| **Burial Grounds – Sacral Buildings** | | | | | | | | | | | |
| Concrete objects | | | | | | | | | | | |
| -pyramid | -limestone | • | + | • | + | + | | • | • | • | • |
| | -granite | + | + | • | + | + | | • | • | • | • |
| | -bricks | + | + | • | + | + | • | | | | • |
| -corridors, shafts, chambers | | | • | | • | + | | • | • | | |
| -tempels | -limestone | • | + | • | + | + | | • | • | • | |
| | -granite, diorite | + | + | • | + | + | | • | • | • | |
| | -basalt | + | + | • | + | + | | • | • | • | |
| | -bricks | + | + | • | + | + | • | | | | • |
| -upward path | -limestone | • | + | • | + | + | | • | • | • | |
| | -granite | + | + | • | + | + | | • | • | • | |
| | -basalt | + | + | • | + | + | | • | • | • | |
| -the Sun Boat | -limestone | • | + | • | + | + | | • | • | • | |
| | -bricks | + | + | • | + | + | • | | | | • |
| -boundary wall | -limestone | • | + | • | + | + | | • | • | • | |
| | -bricks | + | + | • | + | + | • | | | | • |
| Shafts, chambers | | • | + | • | + | + | | • | • | • | |
| | | + | + | • | + | + | • | | | | • |
| -mastabas | -limestone | • | + | | + | + | | • | • | • | |
| | -bricks | + | + | | + | + | • | | | | • |
| Shafts, chambers | | | | | | | | | | | |
| -rock graves | | | • | • | • | + | • | • | • | • | • |

The comprehensive processing is illustrated only on typical examples. In this part of the paper also the close connection between the anomalies of the measured quantities and archaeological objects of different character, age and origin is documented.

The practical contribution of geophysical methods for archaeological studies of prehistoric times, the early historical time and the Middle Ages above all in Moravia and in the space of the Memphis necropolis near Cairo can be divided into several spheres. Above all it is the prospection itself. Archaeogeophysical works started there with the measurement at relatively small areas, where they verified their possibilities. The results with the interpretations proper were verified by archaeological investigation (Bořitov, Budkovice, Hradisko sv.Klimenta near Koryčany, Polešovice, Uherské Hradiště, etc.).Gradually, however, one passed to large area works of the sizes of several hectares whose objective was based above all on the suggestion of suitable sectors of the structure - in Moravia in a great majority of open or fortified settlements, in Egypt of sacral buildings, burial grounds, etc. to further archaeological investigation. These methods represented rationalization in the selection of optimum parts of the studied areas to archaeological excavations proper and the concentration of the attention of the archaeologist, capacities and means on places important from the point of view of settlement, cult or production for the recognition of the station of interest. It was proved that the most important information and results in that sphere were brought by the prospection of settlements (Mikulčice, Pohansko, Olomouc-Slavonín), cult places (Abussir, Al-Ma abda) and also of production sites, particularly metallurgical workshops (Sudice near Boskovice, Habrůvka, Polesí-Olomučany). A somewhat more limited space was that of the study of the prospection of burial grounds. The overall solution is of fundamental importance in finding this type of archaeological monuments in Moravia (Velešovice, Podolí near Brno, Mikulčice, Prušánky, Bohuslavice, Borotice, etc.). In Egyptian archaeology there are more suitable conditions for the above purposes thanks to a greater differentiation between the physical parameters of the objects of interest and the surrounding environment.

Archaeogeophysical investigations gradually looked for further possibilities of their application in which it was passed from the mere form of prospection (the tipping of certain objects or sectors of a station to further investigation, i.e. stations known already from earlier surface investigations or finds and excavations) to obtaining information of a comprehensive character about the station of interest which, due to their lucidity, speed and integrity cannot be replaced by other methods, with the exception of a lengthy and costly archaeological field investigation, either areal or in the form of orientation probing. This concerned e.g. obtaining the whole ground plan situation of the fortification system of fortified settlements and lowland ringwalls (Pohansko, Petrova louka near Strachotín, Spytihněv, Pobedim, Strachotín, Uherské Hradiště - Staré Město, etc.) and hillforts (Diváky, Vedrovice, Blučina, Morkůvky, Kokory, Hluboké Mašůvky) in parts where the fortification was not preserved or has been preserved only in part (filled moats, levelled defence line - vallum fortifications, etc.). Here geophysical measurements bring quick and relatively reliable information, complementing traces of the course of the fortification in the field. On the basis of archaeogeophysical prospection of these settlement types also a new methodology was elaborated, based on a three-stage progress of investigation works. First of all aerial prospection connected with the study of aerial photographs, tipping suitable places of the geophysical measurement. That in turn makes it possible to find the fundamental ground plan diagram of the structure at a certain scale, exactly fixed on the map. The last stage is the verifying probing archaeological investigation. This method found an immediate application and brought positive results mainly in looking for and the study of Neolithic circular structures and other prehistoric objects (Horákov, Křižanovice-Marefy, Němčičky, Rašovice, Šitbořice, Vážany n. Lit., Vedrovice and others) (Hašek - Měřínský 1991: 103-135; Bálek - Hašek 1996: 7-26; Hašek - Kovárník 1996: 57-79), prospection at new line constructions (motorway, high-speed communication) and industrial objects (R 35 Olomouc-Lipník nad Bečvou, traffic detour of Uherské Hradiště, Opava, etc.) (Hašek et al. 1995; 1996; Hašek et al. 1996a). As very prospective it also appears in the case of further settlement structures and formations (Miroslav, Přáslavice near Olomouc). Another sphere where quick and reliable information can be obtained by archaeogeophysical measurement is the prospection of places of medieval exploitation of ore and non-ore raw materials, particularly the mining for precious and ferrous metals (Hory, Svojkovice, Vratíkov near Boskovice). Very good results were also obtained in the application of geophysical methods in the building and historical investigation of medieval architecture. There geophysical works complement the information obtained from the analysis of architecture by further discoveries about the course of today destroyed and non-existent masonry or extinct objects, thus contributing to the recognition of the ground plan structure and the building stages of the studied monuments. Above all this concerns ritual and sacral buildings (Abussir, Al-Ma abda, Újezd near Moravské Budějovice, Předklášteří, Veveří near Veverská Bítýška, Čáslavice near Třebíč, Znojmo, Brno, Šumice,

Doubravník, Jemnice), castles (Bítov, Brtnice, Lelekovice, Sovinec, Rokštejn, Valtice), strongholds (Divice, Daskabát, Tečovice), medieval town and borough centres (Brno, Jihlava, Hlučín, Nový Jičín, Mikulov, Znojmo), etc.. Positive results were also brought by these measurements in the areas of extinct medieval villages, where with their help it is possible to specify the overall ground plan structure, layout and dislocation of the individual objects (Bystřec, Srnávka, Kravín, Mstěnice, Sezimovo Ústí).

The results submitted in the publication, the amount of information utilizable both in archaeological field practice and in theoretical investigation witnesses the variety of the issues solved, evoked by the practical needs of different archaeological institutions, particularly the AS CR, SAS, universities, museums etc., document clearly the irreplaceable position of geophysics in the comprehensive archaeological research.

# 6. CONCLUSION

The investigation of partial information and results obtained by archaeogeophysical prospection above all in Moravia, Slovakia, Germany and Egypt, processed within this publication, can be implemented both in social practice and in further methodological development of applied geophysics. As followed from the preceding chapters, geophysical methods have found full assertion in comprehensive archaeological investigations of stations included in systematic and advance investigations and rescue activities of different institutions. That happened due to the elaboration of the methodology of works and the gradual extension of the application possibilities of geophysical methods within the interdisciplinary co-operation of a number of professional and scientific workers, both from the Archaeological Institute of the Academy of Sciences, Czech Republic, universities, museums, etc., as well as further workers interested in natural historical and technical disciplines, which can be considered a progressive form of the development of this historical branch of science. Thus, a team was formed which was oriented on the comprehensive solution of archaeological issues. As part of this activity, archaeogeophysical prospection was carried out as well as a partial solution of results at a number of stations in Moravia and abroad. Geophysical works, the first of their kind performed within east European countries e.g. under the desert conditions of the Memphis necropolis and near Abussir in Egypt showed besides other things the broad practical application of these methods and their economic advantage in the above archaeological region. On the basis of information obtained by the methods of applied geophysics in the comprehensive research of some old Egyptian monuments near Abussir the Archaeological Institute of the Academy of Sciences, Czech Republic, Brno, was asked for co-operation with foremost institutes abroad, carrying out investigations at different stations of Upper and Lower Egypt, Germany, Austria, Poland, etc..

The number of programs submitted in the monograph and set up for the evaluation and interpretation of the measured data finds a broader application at further geophysical and archaeological institutions. Their importance increases particularly at present, when measurement for the needs of archaeology and/or engineering geology are performed by means of modern digital apparatus, the data being stored in a semiconductor recorder. The interpretation of a great amount of field data can no longer be done by classical methods and thus for their evaluation computer technology is used in 90 % of cases. The above methods of primary processing, qualitative and quantitative interpretation up to the stage of verifying works have been included in the set of standard programs of different organizations dealing with the application of geophysics in archaeology.

The created physico-archaeological models permitted to form a qualitative and spatial idea of the archaeological structure as a whole as well as of the individual parts. These models find immediate application at archaeological institutions and they are a valuable output for the subsequent archaeological investigation, either areal or the verifying probing. Thus the work is made rational and effective both in the methodological, temporal and economic aspects and from the point of view of the number of workers. In the last 10 years the volume of geophysical measurements for purposes of archaeological investigation and prospection, thanks to the results obtained, has increased more than three times.

The processed methodology of archaeogeophysical prospection for the conditions of the Czech Republic and those abroad and the interpretation of the established facts, together with the utilization of modern computer technology, progressive methods of numerical processing and aerial prospection made it possible to turn the attention to those spheres of investigation in which geophysics can complement archaeological information, extend it and thus make it possible to form broader conclusions based on archaeological excavations combined with large-area geophysical measurements. That concerned above all the investigation of fortified settlements, the so-called Neolithic roundels, extinct medieval villages, elements of architecture, above all sacral, but also profane objects, such as castles and strongholds, where geophysical works could be suitably combined with the information of building historical investigations in verifying assumptions and hypothetical solutions obtained by them about the building development and ground plan layouts where relics of overground masonry have no longer been preserved and where its course, on the basis of different indications, was only assumed. There co-operation of geophysics was started not only with archaeologists, but also with the history of art and the study of architecture, its building development and changes (Hašek - Měřínský 1989: 103-151). It is thus possible to quickly and effectively obtain the ground plan situations of different building formations, prehistoric and those of early and culminating Middle Ages which otherwise would be unidentifiable without lengthy and costly investigations.

Indisputable scientific and practical success of applied geophysics in this branch of historical sciences must be developed even in the future, and the share of its activity must always be increased e.g. in a consistent securing of social orders in the sphere of archaeological investigation in the field, further in different forms of of co-operation abroad, etc. The objective must always be the extending application of the scientific recognition aimed at the reconstruction of the historical development of the human society from the Paleolithic up to the Slavonic period and the culminating Middle Ages.

# REFERENCES

Aitken, J. 1958 : Magnetic prospecting the Water Newton Survey. Archeometry I, 1.

Aitken, J. 1959 : Magnetic prospecting. An interim assessment. Antiquity, 33.

Aitken, J. 1961 : Physics and archaeology. Interscience Publishers. London.

Anders, 1971 : Zur geoelektrischen Nachweisbarkeit und Tiefentestimmung von Störkörper im zweischichtigen Halbraum, Neue Bergbautechnik 1, 5-9

Anders, 1972 : Zur metodischen Bassis der geophysikalischen Hohlraumerkundung . Neue Bergbautechnik 2,12-17

Atkinson, R.J.C. 1952 : Méthodes électriques de prospection en archéologie. Paris

Bálek, M., Hašek, V. 1986 :Neue Kreisgrabenanlagen der Kultur mit Mährischer bemalter Keramik in Mähren. Intern. Symp. über die Lengyel-Kultur, Nové Vozokany 1984, 19-26, Nitra-Wien

Bálek, M., Hašek, V. 1996 : Přínos letecké a geofyzikální prospekce pro poznání nových výšinných opevněných sídlišť na Jižní Moravě. Jižní Morava 32, 35, 7-26

Bálek,M., Hašek, V., Měřínský,Z., Segeth,K. 1986 : Metodický přínos kombinace letecké prospekce a geofyzikálních metod při archeologickém výzkumu na Moravě. AR XXXVIII, 550-574

Bálek, M., Havlíček, P. 1987 : Ověření geofyzikálního měření u Divák , okr. Břeclav. PV 1985, 85-86. Brno

Banning, E.J. et al. 1980 : Electrical Conductivity Measurements in the Wadi Tumilat, Egypt, Using an Electromagnetic Surveying Device. Symp. on Archaeometry and Archaeol. Prosp. Paris

Bárta, V. 1971 : Použití geofyzikálních metod při výzkumu zaniklé osady Záblacany, okr. Uherské Hradiště. In : Zaniklé středověké vesnice v ČSSR ve světle archeologických výzkumů. II, 117-124. Uherské Hradiště

Bárta, V. 1973 : Aplikace geofyziky pro archeologické účely v n.p.Geofyzika, závod Praha. Zprávy ČSSA při ČSAV XV, vol. 1-3, 6-7

Bárta, V. et al. 1980 : Použití geofyziky při výzkumu středověkých lokalit v Čechách. In : Sborník referátů 1. celost. konf. "Aplikace geofyz. metod v archeol. a moder. met. terén. výzk. a dokum." Petrov nad Desnou, 9-13. Brno

Bárta, V., Man, O., Mašková,A. 1985 : Aplikace a zpracování nových geoelektrických měření v archeologii. MS Geofond Praha

Bárta, V., Marek, F., Pleslová,E. 1987 : Přehled výsledků geofyzikálního výzkumu a průzkumu archeologických lokalit v Čechách v letech 1983-85. Sbor. 5. celost. konf. Archeológia - Geofyzika - Archeometria, 10-20, Nov, Vozokany

Belcredi, L. : 1993 : Archeologický výzkum kaple svat, Kateřiny a areálu kláštera Porta coeli v Předklášteří u Tišnova. AH 18, 315-343

Bernat, J., Hašek, V. 1973 : Příspěvek k průzkumu podzemních dutin v okolí hradu Veveří. Zprávy ČSSA při ČSAV XV, seš. 1-3, 8-14

Bevan, B.W. 1983 : Electromagnetics for Mapping Buried Earth Features. Journal of Field Archaeol. 10, 47-54

Bílý, M. 1983 : Geofyzikální radiolokační metoda v archeologii. Sbor. 4. celost. konf. Geofyzika a archeologie. 133-138. Liblice

Bláha, V., Chyba, J. 1978 : Metoda velmi dlouhých vln v odporové modifikaci . MS Geofond Praha

Bláha, V. et al. 1973 : Metoda velmi dlouhých vln. MS Geofond Praha

Breiner, S. 1973 : Applications Manual for Portable magnetometers. Sunnyvale

Brizzolari, E. 1975 : Results in a research of buried cavities by resistivity profiles methods in the foundations area a Viaduct. Annales de Geophysique XXIII, 1, 1-12

Buchheim, 1952 : Beitrage zur Theorie der geoelektrischen Aufschlusmethoden. Freiberger Forschungshefte, Reihe C, Heft 6,Berlin

Carabelli, E. 1967 : Ricerca sperimentale dei dispositivi peri adatti alla prospezione elettrica di cavity scoteterrance. Prospezioni Archeologiche, 2, 9-21

Clark, A.J. 1986 : Archealogical Geophysics in Britain. Geophysics 51, 7, 1404-1413

Čepela,P. 1989 : Výsledky geofyzikálních měření na místech zaniklých středověkých skláren. Sbor. 6.celost. konf. Geof. v arch. a moder. met. ter. výzk.a dokum. Gottwaldov 1988

Dachnov,V.N. 1951 : Električeskaja rozvedka neftjanych i gazovych mestorožděnij. Gostoptechizdat

Dey,A. et al. 1975 : Electric field response of two dimensional inhomogenities to unipolar and bipolar electrode configurations. Geophysics 40, Nr. 4, 566-572

Eisler, J., Pejša, J., Preuss, K. 1988 : A digital model of archaeological excavations in Egyptology. Computer and Quantitative Methods in Archaeology 1988, ed. S.P.Q. Rahtz BAR Intern. Series 446, 109-132, Oxford

Faldus, R. et al. 1963 : A study of the electromagnetic field of a magnetic vertical dipole on the model of homogeneous half- space a spherical cavity. Studia geophysica et geodaetica 7, 372-393

Fletcher, R., Powell,M.J.D. 1963 : A rapidly convergent descent method for minimization. Computer Journal, 7, 149-154

Frantov, G.S., Pinkevič, A.A. 1973 : Geofizika v archeologii.- Leningrad

Frohlich, B., Ortner, D.J. 1982 : Excavations of the Early Bronze Age cemetery at Bab edh-Dhra. Jordan 1981. A preliminary report : Dept. of Antiquities 26, 249-267, 491-500

Frohlich, B., Lancaster, W.J. 1986 : Electromagnetic surveying in current Middle Eastern archaeology : Application and evaluation. Geophysics 51, 7, 1414-1425

Fuchs, G.,Hašek,V., Přichystal, A. 1995 : Application of Geophysics in the Research of Ancient Mining in Egypt. Cong. Metallurgy and Mining in Ancient Egypt, Cairo

Gajdoš, V., Tirpák, J. 1989 : Analýza vztahov medzi anomálnym telesom elektrodovým systemom a krokom merania pri odporovom profilovaní. Sbor. 6. celost. konf. Geof. v arch. a moder. met. ter. výzk. a dokum. Gottwaldov 1988

Golcman,F.M., Kalinina,T.B. 1983 : Statističeskaja interpretacija magnitnych i gravitacionnych anomalij. Nedra. Leningrad

Griffin,W.R. 1949 : Rezidual Gravity in theory and practice. Geophysics V, Nr. 14, 39-56

Grodnicki, J. 1977 : Metoda wyznaczanija anomalii ekstremalnych i przyklady jej zastosowania. Biul. Ing. Geof. 1, 58-79

Gruntorád,J., Karous, M. 1972 : Geoelektrické metody průzkumu. I. díl. Praha

Gupta,R.N., Bhattacharya, P.K. 1963 : Unipol methods of electrical profiling. Geophysics 28, Nr. 4, 608-616

Habberjam, G.M. 1969 : The location of spherical cavities using a tripotential technique. Geophysics 34, 780-785

Halíř,J., Hašek, V. 1989 : Magnetometrické modelování v archeologii. Sbor. 6. celost. konf. Geof. v archeol. a moder. met. ter. výzk. a dokum. Gottwaldov 1988

Hašek, V. 1972 : Výpočet křivek elektrického sondování použitím samočinného počítače MINSK-22. Geol. průzkum 14, 11, 333-336, SNTL Praha

Hašek, V. et al. 1975 : Uplatnění geofyzikálních metod při archeologickém výzkumu v Uherském Hradišti - Sadech. MS Geofond Praha

Hašek, V. et al. 1978 : Aplikace geofyzikálních metod v inženýrské, geologii. Geofyzikální průzkum přehradních profilů. MS Geofond Praha

Hašek, V. et al. 1981 : Geofyzikální radiolokační metoda - vývoj geofyzikálních metod pro vyhledávání dutin a jiných připovrchových nehomogenit. MS Geofond Praha

Hašek ,V. et al. 1982a : Výzkum komplexu geofyzikálních metod pro hnědouheln. pánve. MS Geofond Praha

Hašek, V. et al. 1983 : Podíl geofyzikálních metod při přípravě terénního archeologického výzkumu - etapa 1982. MS Geofond Praha

Hašek, V. et al. 1983a : Výzkum komplexu geofyzikálních metod pro podkrušnohorské pánevní oblasti - etapa 1982. MS Geofond Praha

Hašek, V. et al. 1984a : Výzkum geofyzikálních metod pro podkrušnohorské pánevní oblasti - etapa 1983. MS Geofond Praha

Hašek, V. et al. 1985 : Podíl geofyzikálních metod při přípravě terénního archeologického výzkumu - etapa 1984. MS Geofond Praha

Hašek, V. et al. 1986 : Výzkum geofyzikálních metod pro podrušnohorské pánevní oblasti. MS Geofond Praha

Hašek, V. et al. 1988 : Geofyzikální příprava těžby hnědého uhlí. MS Geofond Praha

Hašek, V. 1988a : Geophysical Data processing systems in Czechoslovak archaeology. Computer and Quantitative Methods in Archaeology, ed. S.P.Q. Rahtz. BAR Intern. series 446, 195-200. Oxford

Hašek, V. 1995 : Zpráva o činnosti odborné pracovní skupiny pro přírodovědecké metody v archeologii za r. 1994. Zprávy ČAS 48, 3-4

Hašek, V., 1996 : Zpráva o činnosti odborné pracovní skupiny pro přírodovědecké metody v archeologii za r. 1995. Zprávy ČAS 49,3-4

Hašek, V. 1997 : Zpráva o činnosti odborné pracovní skupiny pro přírodovědecké metody v archeologii za r. 1996. Zprávy ČAS 50, 3-4

Hašek, V., Horák,T. B., Obr, F. 1979 : Uplatnění přírodovědných metod při archeologickém výzkumu hutnického střediska v Sudicích. Sbor. Okresního vlastivědného muzea v Blansku X, 46-60

Hašek, V., Kovárník, J. 1996 : Letecká geofyzikální prospekce při výzkumu pravěkých kruhových příkopů na Moravě. SPFFBU, M 1, 57-79

Hašek, V., Kovárník, J. 1996a : Geofyzika v moravské středověké archeologii. Muz. a vlast. práce 2/96, 65-88

Hašek, V., Ludikovský, K. 1977 : Interdisciplinární racionalizační brigáda, vyšší stupeň mezioborové spolupráce v geofyzikálním a archeologickém výzkumu. Geologický průzkum XIX, č. 4, 117-119

Hašek, V., Ludikovský, K. 1977a : Některé výsledky prací moravské sekce interdisciplinární brigády v roce 1976. Zprávy ČSSA při ČSAV XIX, seš. 4-5, 108-115

Hašek, V., Ludikovský, K., Obr, F. 1979 : Geofyzikální a geochemický výzkum fortifikačního systemu na slovanském hradisku v Pobedimi. Geol. průzkum XXI, 2, 45-48

Hašek, V., Měřínský, Z. 1987 : Podíl geofyziky při archeologických výzkumech na Moravě v letech 1983-1985. In : Acta interdisciplinaria archaeologica, 102-140. Nitra

Hašek, V., Měřínský, Z. 1987a : Geofyzikální příprava terénního archeologického výzkumu, etapa 1987. MS Geofond Praha

Hašek, V., Měřínský, Z. 1989 : Podíl IRB při archeologickém výzkumu a archeogeofyzikální prospekci na Moravě v letech 1986 - 88. Sbor. 6. celost. konf. Geof. v archeol. a moder. met. ter. výzk. a dokum. Gottwaldov 1988, 103-151

Hašek, V., Měřínský,Z. 1989a : Magnetometric modelling in Archaeology. Com. 4. Comp. and Quantitative Methods in archaeology. BAR, Mogilany

Hašek, V., Měřínský, Z. et al. 1991 : Archeogeofyzikální prospekce na Moravě. MVS Brno

Hašek, V., Měřínský, Z., Págo, L. 1983 : Interdisziplinare Rationalisierungsbrigade (IRB) - Arbeitsergebnisse des mährischen Teiles für das Jahr 1981. PV 1981, 81-82. Brno

Hašek, V., Měřínský, Z.,Págo, L. 1984 : Interdisziplinare Rationalisierungsbrigade (IRB) -Arbeitsergebnisse im Jahre 1982. PV 1982, 102-104. Brno

Hašek,V., Měřínský, Z., Págo,L. 1985 : Interdisziplinare Rationalisierungsbrigade (IRB) - Arbeitsergebnisse im Jahre 1983. PV 1983, 115-116. Brno

Hašek, V., Měřínský, Z, Págo, L. 1987 : Interdisziplinare Rationalisierungsbrigade (IRB) - Arbeitsergebnisse für Jahr 1984. PV 1984, 93-94. Brno

Hašek, V., Měřínský, Z., Págo, L.1987a : Interdisziplinare Rationalisierungsbrigade (IRB) - Arbeitsergebnisse im Jahre 1985, PV 1985, 89-90. Brno

Hašek, V., Měřínský, Z., Segeth, K. 1990 : New Trends in Processing and Interpretation of Geophysical Data in Czechoslovak Archaeology. Communication in Archaeology a global view of the impact of information technology, 27-34

Hašek, V., Měřínský, Z.., Unger, J., Vignatiová, J. 1983 : Výsledky geofyziky v archeologickém výzkumu a průzkumu na Moravě v letech 1979-1982 a jejich metodický přínos. In : Geofyzika a archeologie, 141-153. Praha

Hašek.V., Obr, F.,Přichystal, A., Verner,M. 1986 : Application of geological and geophysical methods in archaeological research at Abusir. Sbor. geol. Věd, řada HIG 18, 149-197. Praha

Hašek, V.,Obr,F., Verner, M. 1988 :Application of geological and geophysical methods in archaeological investigations of ancient Egyptian Remnants at Abusir. Przeglad Archeologiczny 35, 5-47

Hašek, V., Petrová, H., Segeth, K. 1994 : Graphic representation methods in archaeological prospection in Czechoslovakia. Comp. and Quantit. Methods in Archaeology CAA 92, Aarhus University Press ,63-66

Hašek, V., Rössler-Köhler, U. 1996 : Geophysical and archaeological Investigation of Ancient Town at Al Maabda in middle Egypt. Przeglad archeologiczny (in print)

Hašek,V., Segeth,K., Vencálek,R. 1990 : Některé nové možnosti aplikace numerického zpracování geofyzikálních dat v archeologii. Sbor. 6. celost. konf. Geofyzika v archeol. a Moder. met. ter. výzk. a dokum. Gottwaldov 1988

Hašek, V., Uhlík, J. 1991 : Geofizičeskoe isledovanie těl pročnych porod v pokrovnych otloženijach pačkach burych uglej severočeskogo bassejna. Sbor. geol. věd G, řada HIG, 19, 155-192

Hašek, V., Unger, J. 1994 : Archöogeophysikalische Prospektion der historischen unterirdischer Röume in der Tschechischen Republik. Der Erdstall 20, 30-43

Hašek, V., Unger, J., Záhora, R. 1997 : Archäologische prospektion mit Georadar in Mähren. Beitr.für Mittelalterarchäologie in Österreich, 13, 23-39.

Hašek, V., Vencálek, R. 1989 : Method for processing digital geophysical data in Archaeology. Computer and Quantitative Methods in Archaeology, ed. J. Richards, BAR Intern. series, YORK, 179-192

Hašek, V., Vencálek, R. 1991 :Normalized Gradient Method of Quantitative Interpretation of Geophysical Data in Archaeology. Comp. and Quantit. Methods in Archaeology 1990, BAR Intern. Series Oxford

Hašek, V., Veselý, I., Woznica, L. 1981 : Uplatnění povrchov, geofyziky při inženýrskogeologickém průzkumu přehradních profilů, Sbor. geol. Věd. řada HIG, 15, 83-126, Praha

Hvožďara, M., Schlosser, B. 1985 : Anomalia tellurického a termického pola vyvolaná prítomnsťou dvojrozmerného telesa v homogénnom polpriestore. Contribution of Geoph. Inst. SAS, 15, 35-49

Hvožďara, M.,Tirpák, J. 1987 : Modelovanie elektrických polí v prítomnosti dvojrozmerných nevodičov pre ciele odporového profilovania v archeologii. Sl. Arch. 1, XXXV-1, 165-188

Chyba, J. 1981 : Zpracování výsledků elektrické sondáže metodou nejmenších čtverců. MS Geofond Praha

Jain, S.C. 1974 : Theoretical broadside resistivity profiles over an outcropping dyke. Geophysical Prospecting 22, 445-457

Jones, F.W., Pascal, L.J: 1971: A general computer program to determine the perturbation of alternating electric currents in a two dimensional model of a region of uniform with an emtedet.

Karous, M., 1982 : Odporové a fázové křivky metody velmi dlouhých vln v odporové variantě. Geol. průzkum 24, 3, 77-79. SNTL Praha

Karousová, O. 1979 : Dekonvoluce profilových křivek T (dipl. práce). MS PřF UK Praha

Kenyo, B .I. 1977 : Ground - penetrating radar and its application to a historical archaeological site. Archeol. 11, 48-55

Kowalik, W. S., Glenn, W. E. 1987 : Image processing of aeromagnetic data and integration with Landsat images for improved structural interpretation. Geophysics 52, 875-884

Kumar, R. 1973 : Resistivity type curves over outcropping vertical dyke I. Geophysical Prospecting 21, 560-578

Kumar, R. 1973a : Resistivity type curves over outcropping vertical dyke II. Geophysical Prospecting 21, 615-625

Kyono, T. 1950 : Theoretical study of the ground resistivity methods of electrical prospecting, Kyoto-Univ. Faculty Eng. M.V. 12, 29-59

Laksahmann, J. 1963 : Feststellung von unterirdischen Hohlräumen mit elektrischen und Gravimetrischen methoden. "Z Sols Soils", 2, 9-15. Paris

Latka, R. 1966 : Modellrechnungen zur Induktion in leicht föhigen Nortergrund . Zeitschrift für Geophysik 32, 512-517

Lerici, C. M. 1955 : Prospezioni Archaologiche. Revista geofisica applicata, Anno XVI, 1-2

Lerici, C.M. 1960 : I nuovi metodi di prospezione archeologica alla scoperta delle civilta sepolte. Lerici editori. Milano

Ljubimov, G. A., Ljubimov, A. A. 1983 : Vyčislitělnyje schemy dlja sglaživanija, interpolacii i opredelenija vyšších proizvodnych geofizičeskich polej. Razv. geof. 130-136

Logačev, A.A., Zacharov, V.P. 1979 : Magnitorazvedka. Moskva

Lösch, W., Militzer, H., Rösler, R.1979 : Zur geophysikalischen Hohlraumortung mittels geoelektrischer Wiederstands-methoden. Freiberger Forschungshefte 341, 53-126

Linington, R.E. 1969 : The prospecting campaign undertaken in Czechoslovakia in July-August 1968. Prospezioni archeologiche Fondazione Lerici 4, 131-138

Linington, R.E. 1970 : Prospecting methods in archaeology. AR XXIV, 169-194. Praha

Ludikovský, K. 1978 : Die Fragen der archeologischen Komplexfeldforschungen in Mähren. In : Intern. symp. mechanization of the archeolog. Fieldworks Archaeol. Baltica III, 79-103. Lodž

Ludikovský, K., Hašek, V., Obr, F. 1978 : Geofyzikální výzkum příčného valu na slovanském hradisku v Pobedimi. Slovenská arche

Malina, J. 1976 : Archeologie, jak a proč ? Mikulov. Břeclav

Mc Neil, J. 1980 : Electromagnetic terrain conductivity measurement at low induction numbers. Geonics Limited TN-6, 5-14

Marek, F., Plesl, E. 1978 : Geophysikalische Prospektion in Böhmen. Archeol. Baltica 3, 104-131

Marek, F., Pleslová, E. 1977 : Geofyzikální prospekce na archeologické lokalitě Makotřasy , okr. Kladno. Zprávy ČSSA při ČSAV XIX, seš. 4-5, 98-101

Mareš, S. et al. 1979 : Úvod do užité geofyziky. Praha

Mareš, S. et al. 1983 : Geofyzikální metody v hydrogeologii a inženýrské geologii. Praha

Marquardt, D.W. 1963 : An algorithm for least squares estimation of nonlinear parameters. Journal Soc. Industr. Appl. Math. 11, 431-441

Mašín, J., Válek, R. 1963 : Přehled užité geofyziky pro geology. 293-294. Praha

Meyer, I. R. 1970 : Theoretical and computational aspects of non-linear regression in "Non-linear programming", edited by J.B. Rosen. New York and London. Academic Press

Militzer, H., Rösler, R., Lösch, W. 1979 : Theoretical and Experimental Investigations Methods. Geophysical Prospecting 27, 640-652

Moffatt, D.L. 1974 : Subsurface video pulse radar. Proc. Rng. Found. Conf. on Subsurface Expl. for Undergroud Excav. and Heavy Constr. New England College 16

Moločnov, G.V., Balobajev, V.T. 1958 : Provodjaščie telo v elektromagnitnom pole vertikalnogo magnitnogo dipolja. Voprosy geofiziki, Uč. zap. LGU 249, Seria geof. a geol. nauk 10, 80-89

Money, R.M. 1974 : Continuous subsurface profiling by impulse radar. Proc. of an Eng. Found. Conf. on Subsurface Expl. for Underground Excav. and Heavy Constr. New England College, 11-16, 213-232

Müller, K., Müllerová, J. 1976 : Použití geofyzikálních metod pro sledování deformací povrchu vlivem hornické a stavební činnosti. Sborník II. konference "Aplikace geofyzikálních metod v hydrogeologii a inženýrské geologii". Brno

Müller, K., Okál, M., Hofrichterová, L. 1985 : Základy hornické geofyziky. SNTL Praha

Nesterov, L. et al. 1938 : Kurs elektrorazvedki. Leningrad

Nikitin, A.A. 1986 : Teoretičeskie osnovy obrabotki geofyzičeskoj informacii. Nedra. Moskva

Novotný, V. 1973 : Uzemnění a jeho měření. SNTL Praha

Odstrčil,J. 1985 : Gravimetrický programový system pro minipočítače. MS Geofyzika s.p. Brno

Odstrčil, J. 1989 : Poznámky k separaci gravimetrického a geomagnetického pole. Sbor. 6. celost. konf. Geof. a arch. a moder. met. ter. výzk. a dokum. Gottwaldov 1988

Parasnis, D.S. 1965 : Theory and practice of potential and resistivity prospecting using linear current electrodes. Geoexploration 3, 3-69

Pašteka, V. 1977 : Transformácia pola v magnetometrii. Banické listy, BÚ SAV, 80-88

Podborský, V. 1977 : Dějiny pravěku. FF UJEP Brno

Pretlová, V. 1976 : Bicubic spline smoothing of two-dimensional geophysical data. Studia geophysica et geodaetica 20, 168-177

Rasmussen, R., Pedersen, L.B. 1979 : End correction impotential field modelling. Geph. Prosp. 27, 749-760

Scollar, I. 1965 : A contribution to magnetic prospecting in Archaeology. Band 15, 21-92. Köln-Graz

Sikorskij, V.A. 1979 : Komplexnaja interpretacija geofizičeskich dannych na osnove koreljacionnoj samonastrojki. ANSSSR Geologija a geofizika 10, 120-126

Švancara, J.,Halíř, J. 1986 : Obrácená gravimetrická Úloha pro 2 1/2 D struktury. Celost. sem. "Probl. souč. gravim". Sbor. ČSAV a Geofyzika s.p. Brno

Tabbagh, A. 1974 : Methods de prospection electromagnetique applicables aux issuees archeologiques. Archaeophys. 5, 350-437

Tarchov, A.G. et al. 1982 : Komplexsirovanie geofizičeskich metodov. Moskva

Tirpák, J., 1977 : Prieskum niektorých archeologických lokalit na Slovensku. Zprávy ČSSA při ČSAV XIX, seš. 4-5, 120-122

Tirpák, J. 1984 : Modelovanie elektrických polí v prítomnosti dvojrozmerných nevodičov pre ciele odporového profilovania v archeológii. Rkp. kand. disertační práce SAV Bratislava

Tomek, Č. 1975 : Problémy separace potenciálních polí. MS Geofond Praha

Töepfer, K. 1969 : Die Ortung von Störkörpern mit dem Schlumberger Messverfahren. Arch. met. Geoph. Bickl. A 18, 191-220

Uhlík, J. 1968 : Anwendung der Methode der Potentialverhältnisse für die Ortung von Hohlräumen. Uhlí 10, Nr. 12, 457-460

Vaughan, C.J. 1986 : Ground-penetrating radar surveys in archaeological investigations. Geophysics 51, 7, 594-604

Verner, M., Hašek, V. 1981 : Die Anwendung geophysikalischer Methoden bei der archäologischen Forschung in Abusir. Zeitschr. für Ägypt. Sprache 108, 68-84. Berlin

Věšev, A.V. 1980 : Elektroprofilirovanije na postojannom toke. Leningrad

Wait, D.R. 1954 : Matual Coupling of loops lying on the ground. Geophysics 19, 290

Wait, D,R. 1955 : Matual electromagnetic coupling of loops over a homogeneous ground. Geophysics 20, 630

Weymouth, J. W. 1986 : Archaeological site surveying program at the Univ. Nebraska. Geophysics 51, 7, 538-552

Withe, R. 1970 : Gutachten über geoelektrische Messungen im Tagebau Klettwitz. VEB Geophysik Leipzig

Wynn, J. C. 1986 : Archaeological prospection : An Introduction to the Special Issue. Geophysics 51, 533-537

Záhora, R. 1979 : Potenciální možnosti geofyzikální radiolokační metody v podmínkách SHD. MS Geofond Praha

Záhora, R. 1989 : Využitelnost přímého bezkontaktního měření zdánliv, vodivosti pro Účely archeologické prospekce. Sbor. 6. celost. konf. Geof. v archeol. a Moder. met. ter. výzk. a dokum. Gottwaldov 1988

Zelenka, P. 1985 : Geofyzikální průzkum na archeologických lokalitách Diváky a Mikulčice (Diplomová práce). MS PřF UK Praha

1. Magnetometric measurement with gradiometer PMG-1 (photo W. Steeger)

2. Working with conductometer KD-2 (photo W. Steeger)

3. Soil radar RAMAC/GPR (photo W. Steeger)

4. Pálava, district of Břeclav: Palaeolithic station of Pavlov in background, primaeval small fort on tableland in foreground
(photo J. Kovárník)

5. Rašovice, district of Vyškov: Aerial photo of Neolithic circular enclosure (photo J. Kovárník)

6. Vážany on Litava, district of Vyškov: Aerial photo of semicircular structure (photo J. Kovárník)

7. Vedrovice, district of Znojmo: Circular ditch with Moravian painted point-shaped pottery (photo V. Hašek)

8. Vedrovice, district of Znojmo: Uncovered objects with linear pottery inside circular enclosure with MMK (photo V. Hašek)

9. Mušov 'Na pískách', district of Břeclav: Aerial photo of a Roman marching camp with objects of habitation (photo J. Kovárník)

10. Rokštejn, district ofJihlava: Overall view of relics of a castle (photo Z. Měřínský)

11. Valtice, district of Břeclav: Aerial photo of Baroque castle, location of Gothic castle in background (photo J. Kovárník)

12. Božice, district of Znojmo: Aerial photo of a circular structure (photo J. Kovárník)

13. Božice, district of Znojmo: Uncovered medieval ditch with Slavonic graves (photo V. Hašek)

14. Znojmo: Overall view of Saint Nicholas Church (photo J. Kovárník)

15. Předklášteří near Tišnov, district of Brno-country: Uncovered monastery chapel of Saint Catarina (photo L. Belcredi)

16. Velešovice, district of Brno-country: Grave of Cord-Pottery-Culture (photo M. Bálek)

17. Egypt: Overall view of Abusir burial site (photo ČSEÚKU archives)

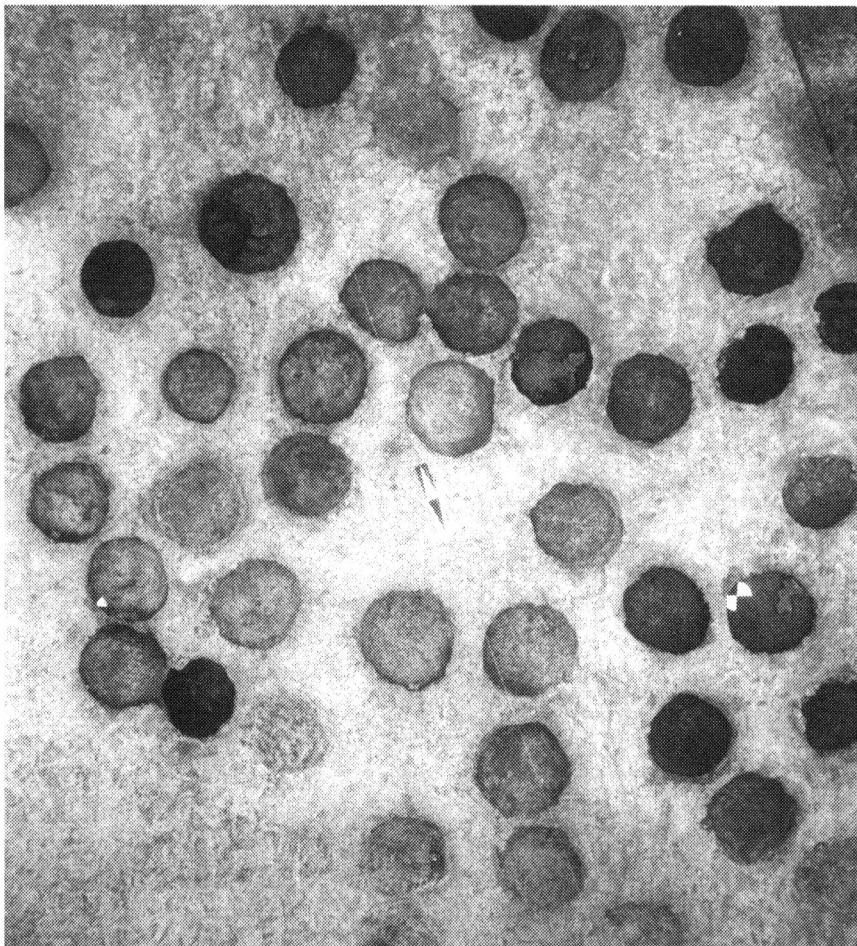

18. Sudice near Boskovice, district of Blansko: Hearth of iron furnace from Early Roman period

www.ingramcontent.com/pod-product-compliance
Lightning Source LLC
Chambersburg PA
CBHW051301270326
41926CB00030B/4689